M000167443

Quasicrystals and geometry brings together for the first time the many strands of contemporary research in quasicrystal geometry and weaves them into a coherent whole. The author describes the historical and scientific context of this work, and carefully explains what has been proved and what is conjectured. This, together with a bibliography of over 250 references, provides a solid background for further study.

The discovery in 1984 of crystals with 'forbidden' symmetry posed fascinating and challenging problems in many fields of mathematics, as well as in the solid state sciences. By demonstrating that 'order' need not be synonymous with periodicity, it raised the question of what we mean by 'order', and how orderliness in a geometric structure is reflected in measures of order such as diffraction spectra. Increasingly, mathematicians and physicists are becoming intrigued by the quasicrystal phenomenon, and the result has been an exponential growth in the literature on the geometry of diffraction patterns, the behavior of the Fibonacci and other nonperiodic sequences, and the fascinating properties of the Penrose tilings and their many relatives.

This first-ever details account of quasicrystal geometry will be of great value to mathematicians at all levels with an interest in quasicrystals and geometry, and will also be of interest to graduate students and researchers in solid state physics, crystallography and materials science.

QUASICRYSTALS AND GEOMETRY

"Aperiodic Penrose α", by the noted sculptor/mathematician Helaman Ferguson, was commissioned by Marjorie Senechal, who worked closely with him on its design. Completed in 1995, it stands in the lobby of the Clark Science Center at Smith College. The sculpture, of carnelian granite, rests on a base of black granite Penrose tiles. The articulated torus encodes the composition matrix of the tiling.

QUASICRYSTALS AND GEOMETRY

MARJORIE SENECHAL

Louise Wolff Kahn Professor of Mathematics
Smith College, Northampton, Massachusetts

CAMBRIDGE
UNIVERSITY PRESS

Published by the Press Syndicate of the University of Cambridge
The Pitt Building, Trumpington Street, Cambridge CB2 1RP
40 West 20th Street, New York, NY 10011-4211, USA
10 Samford Road, Oakleigh, Melbourne 3166, Australia

© Cambridge University Press 1995

First published 1995
First paperback edition (with corrections) 1996

Printed in Great Britain at the University Press, Cambridge

A catalogue record for this book is available from the British Library

Library of Contgress cataloguing in publication data

Senechal, Marjorie.
Quasicrystals and geometry/Marjorie Senechal
p. cm.
Includes bibliographical references.
ISBN 0-521-37259-3
1. Quasicrystals. 2. Crystallography, Mathematical. I. Title.
QD926.S46 1995
541'.81–dc20 94-17063 CIP

ISBN 0 521 37259 3   hardback
ISBN 0 521 57541 9   paperback

Contents

Preface

It was evident almost immediately after the November 1984 announcement of the discovery of crystals with icosahedral symmetry that new
areas of research had been opened in mathematics as well as in solid state
science. For nearly 200 years it had been axiomatic that the internal
structure of a crystal was periodic, like a three-dimensional wallpaper
pattern. Together with this axiom, generations of students had learned its
corollary: icosahedral symmetry is incompatible with periodicity and is
therefore impossible for crystals. Over the years, an elegant and far-
reaching mathematical theory had been developed to interpret these
'facts'. But suddenly – in the words of the poet W. B. Yeats – *all is
changed, changed utterly*.

A long-planned workshop on mathematical crystallography held in
France two months after the announcement quickly evolved into a
seminar on the possible geometry of the puzzling new structures
(Senechal, 1985). It was widely agreed that many fundamental and
fascinating problems were waiting to be solved. These problems included
such broad questions as: what are the appropriate mathematical models
for crystal structure if we no longer define a crystal to be a solid with a
periodic atomic pattern? What geometric properties must a point set or
tiling model have in order for its diffraction patterns to show sharp, bright
spots? How can such point sets and tilings be generated? How should they
be classified? Indeed, the discovery called for a reexamination of the
mathematical foundations of geometrical crystallography.

Today, after nearly a decade of intense research in laboratories all over
the world, many different kinds of aperiodic crystal have been discovered,
and a great deal is known about their structure and properties. But such is
their complexity that, to date, only one structure – a 'decagonal', not an

'icosahedral' quasicrystal –is widely agreed to have been 'solved' (Steurer and Kuo, 1990).

The situtation is much the same with the mathematical problems that these 'generalized' crystals suggest. Although the number of new and important results is increasing rapidly, there is still no agrement on how aperiodic crystals should be modeled mathematically, and most of the major problems raised by any of the proposed models are still open. This state of affairs might suggest that a book on the subject is premature, but I think the opposite is in fact the case: a book is overdue. This is the stage in which the field is most exciting, precisely because it is in a state of flux and has so many fascinating and challenging unsolved problems. A book now that explains what is not known, rather than a book later that codifies what is, might make a contribution by attracting people to work on these problems.

This book is not about *real* crystals. Two excellent volumes that discuss the real materials and the efforts of physicists to model their properties are now available: *Quasicrystals: the state of the art*, edited by DiVincenzo and Steinhardt (1991) and *Quasicrystals: a primer*, by Janot (1993). We will focus instead on the mathematical concepts and problems that appear to be important for a reconstructed mathematical crystallography.

Mathematical crystallography is, above all, mathematics, but one hopes that it has something to say about crystals too. Years ago a senior crystallographer, Helen Megaw, thought it advisable to remind me that 'a crystal is not a group'. Today it is equally important for mathematicians to remember that a crystal is not a point set or a tiling, either. Crystal geometry is the resultant of chemical and physical forces and processes and no account of it that does not treat its dynamical aspects can be considered complete. But geometry – together with a generalized notion of symmetry – will probably continue to be the basis for crystal classification, and it is the classification problem with which this book is chiefly concerned. Just as groups do help scientists to understand the structures of periodic crystals, the subjects discussed in this book, when properly developed, should be useful for understanding quasicrystals and other 'disordered' solid materials.

The classification of aperiodic crystals requires a larger and deeper mathematical toolbag than was needed for the periodic case, for which group theory and the elements of tiling theory nearly sufficed. Some of these tools were new to me, and although I have enjoyed the adventure of learning how to use them, I am also aware that I may have made errors or am ignorant of the relevant literature. I will be grateful for any criticisms, comments, and suggestions: the adventure continues.

One of the most difficult problems that confronted me in writing this book was deciding what to put in and what to leave out. I have chosen to focus on two central problems, the problems of long-range order and self-assembly. Thus the book deals primarily with the problem of relating the geometry of discrete point sets to the diffraction spectra of functions associated to them, and with the emerging theory of aperiodic tilings. Within these broad categories, my main selection principle has necessarily been my own taste and judgement. In some cases, in order to exhibit connections among the various parts of the problem, I have omitted contributions whose connections I could not forge to my own satisfaction. Finally, I have tried to keep the exposition on an intermediate mathematical level, and to make it relatively self-contained and coherent, so that the book will be accessible to a broad audience of solid-state scientists, mathematicians, students, and the scientifically literate. Unfortunately, this has meant that the theory of generalized functions has had to be relegated to a brief appendix, and interesting work involving measure theory, ergodic theory, homotopy theory, Lorenzian geometry, multifractal analysis, and other advanced topics, is barely mentioned if at all.

A word about rigor. To convey understanding, the writer of mathematics must find a way to balance many modes of thought: linguistic and symbolic, algebraic and geometric, logical and intuitive. The choices I have made reflect both my own ways of understanding the material and the responses of the varied audiences to whom I have presented it. In addition, to try to describe a subject in such a rapid state of development – and change – as this one, in a formal style would be as inappropriate as it would be impossible: instead of opening up the subject, it would build a fence around it. The reader should understand that many definitions are necessarily provisional and many proofs are sketchy. I have included at least outlines of proofs of (nonobvious) propositions and theorems whenever the following three conditions could be met: (i) a proof, or at least a convincing outline of a proof, exists, (ii) the proof sheds some light on the subject matter, and (iii) the proof could be explained without recourse to mathematical techniques hopelessly far beyond the scope of the book.

Many people have helped to shape this book. I am grateful to my students at Smith College, whose excellent questions and stubborn disbelief forced me to clarify concepts and refine my arguments. I am also grateful to the many colleagues with whom I have discussed some or all of the issues treated here: N. G. de Bruijn, Ludwig Danzer, Claude Godrèche, Louis Michel, Charles Radin, Doris Schattschneider, Rebekka

Struik, Jean E. Taylor, and others too numerous to mention. Special thanks are due to N. G. de Bruijn, H. S. M. Coxeter, Claude Godrèche, Branko Grünbaum, Mark Heald, Bert Hof, Michel Mendes France, Le Tu Quoc Thang, Louis Michel, Robert Moody, Doris Schattschneider, Christophe Schneider, Joshua Socolar, and Kenneth Stolarsky, who read various drafts of the manuscript (in some cases more than one); their astute comments and helpful suggestions are greatly appreciated. All errors remaining are, unfortunately, my own.

The beauty of crystals has always been part of their fascination for the general public and scientists alike. The nineteenth-century hypothesis that these austere geometric forms are reflections of periodic atomic patterns transformed crystallography into a subject whose aesthetic and scientific aspects were inseparable: each informed the other. Today we are finding that the 'atomic' and diffraction patterns associated with aperiodic crystals are even more beautiful – and, more importantly, even more instructive – than in the periodic case. Thus trying to convey these patterns visually was at least as high a priority for me as trying to express their properties through words and formulas. It is a pleasure to thank the Geometry Center at the University of Minnesota in Minneapolis for a month-long Research Professorship during which I was able to use their superb facilities and software and benefit from the support and guidance of their expert staff. I am especially grateful to the Center's computer graphics wizards Stuart Levy and Eugenio Durand, who wrote the special programs 'fourier' and 'QuasiTiler' to help create the diffraction patterns and other elaborate illustrations. (Most of the line drawings in this book were made with the help of *Mathematica* and *Geometer's Sketchpad*.)

I would also like to thank Simon Capelin, Senior Physics Editor at Cambridge University Press, for his continual encouragement and remarkable patience, and the Institut des Hautes Etudes Scientifiques, in Bures-sur-Yvette, France, for providing the perfect working conditions for completing the manuscript.

Although she did not help in any way, my sister Norma has asked me to thank her, too. (Thanks, sissy!)

This book is dedicated to the memory of David Harker (1906–1990) and Rolph Schwarzenberger (1936–1991). David was a superb crystallographer with a fine appreciation for mathematics; Rolph was a superb mathematician with a fine appreciation for crystallography. Both of them made important contributions to mathematical crystallography. Our professional paths first crossed many years ago through a common interest in color symmetry, and gradually developed into the kind of collegial friendship that provides encouragement and support and makes the scientific life a privilege and a joy. It is our loss that they are not here to help develop this exciting new chapter in our field.

Marjorie Senechal
Northampton, Massachusetts
email address: senechal@minkowski.smith.edu.

1

Past as prologue

The history of its development ... is the history of an attempt to express geometrically the physical properties of crystals, and at each stage of the process an appeal to their known morphological properties has driven the geometer to widen the scope of his inquiry and to enlarge his definition of homogeneity.

(W. Barlow and H. Miers, 1901)

1.1 Demise of a paradigm

Ten years ago, the world of solid state science was startled by the announcement of the discovery of 'a metallic phase with long-range orientational order and no translational symmetry' (Shechtman, Blech, Gratias, and Cahn, 1984) (see Figure 1.1). The material in question was an alloy of aluminum and manganese, produced by Shechtman from a melt by a rapid cooling technique. Its diffraction images showed icosahedral symmetry, long believed to be impossible for matter in the crystalline state. After all, a crystalline structure was *by definition* a periodically repeating pattern, and periodicity is another name for translational symmetry, the strictest form of long-range order. In the absence of evidence to the contrary, and in the presence of long-standing belief, it was widely assumed that the reverse must also be true: long-range order is a synonym for translational symmetry. Translational symmetry and icosahedral symmetry are incompatible (see below). Thus the new metallic phase was very puzzling indeed, since a sharp, clear diffraction pattern is evidence of long-range order. Either the symmetry in the diffraction pattern was an artifact of some sort, or something new and strange had been discovered.

We need not review here the spirited debates that immediately ensued; they have been recounted elsewhere (La Brecque, 1987) and in any case have long since quieted down. Unlike some other famous examples of 'headline physics', this phenomenon did not fade away on close inspection. Just the opposite: not only was this 'phase' quickly reproduced in laboratories throughout the world, but many other examples of 'noncrystalline crystals' were found as well. It soon became clear that aperiodic metallic phases are not rarities but, on the contrary, are very widespread. (As soon as Galileo announced that he had seen

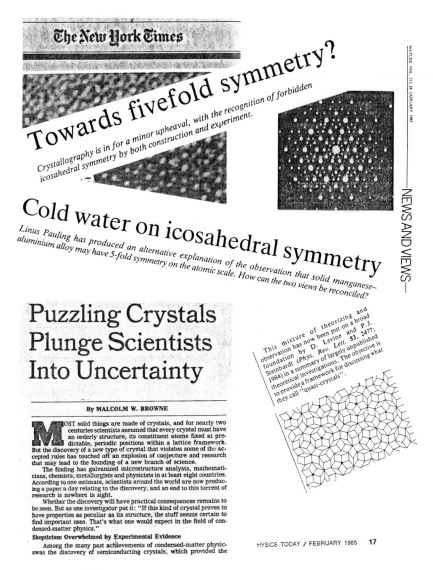

Fig. 1.1 Headlines from the quasicrystal controversy.

mountains and valleys on the moon, everyone else was suddenly able to see them too, although only the day before they had known beyond question that the moon was a perfect sphere (Edgerton, 1984).)

With hindsight we can see that the postulate of periodicity had been weakening for a long time. Within a few years of the 1912 discovery of x-ray diffraction, simple mineral crystal structures had been successfully determined and the scope of crystallography was already being broadened to include the study of the much less orderly structures of fibers and proteins. Since then, the scope of crystallographic inquiry has widened ever more rapidly with the invention of instruments with ever higher resolving power. By 1984 crystallography included the study of the geometry of amorphous structures, complex biological molecules, liquid crystals, magnetic structures, and incommensurate crystals. Even some of the methods used today to study aperiodic crystals had already been developed: P. M. de Wolff introduced four-dimensional space groups to interpret the diffraction patterns of 'modulated' crystals (de Wolff and Aalst, 1972), and his colleagues developed the theoretical framework for this interpretation (Janner and Janssen, 1979). But however much the periodic paradigm had had to be expanded or distorted to accommodate such complex and unruly structures, the laws of crystal *symmetry* had stayed intact. Thus none of these prior developments had the impact of the brief announcement by Shechtman and his colleagues.

Perhaps in no branch of science have the technical and the aesthetic been as closely linked as in crystallography. The shock produced by the brief announcement was due not to the visible absence of translational symmetry but to the visible presence of icosahedral symmetry, evident in the accompanying diffraction photographs. The regular icosahedron and its dual, the regular pentagonal dodecahedron, have five-fold, three-fold, and two-fold rotation axes, as does their close relative, the rhombic triacontahedron (Figure 1.2); the diffraction patterns of the new AlMg phase (Figure 1.3) showed ten-fold, six-fold, and two-fold rotational symmetry. (Diffraction patterns are always centrosymmetric – for the reasons, see Chapter 3.)

A fundamental 'law' of crystallography declares icosahedral symmetry to be impossible in the crystal kingdom. This law, based on the long-established hypothesis of periodicity and taught in every textbook, had been confirmed in all known instances: instructors delighted in pointing out how examples of *apparent* icosahedral symmetry (dodecahedral pyrite, 'fivelings' of silver or gold) could be shown to be irregular or an artifact of twinning (Figure 1.4). (These are some of the phenomena that many people first thought might explain the 'apparent' icosahedral symmetry of quasicrystals.)

A pattern is periodic if it can be subdivided into a countable infinity of congruent regions, called fundamental regions (with respect to translations), such that:

(i) two regions have no common interior points, and the union of the regions is the pattern itself;

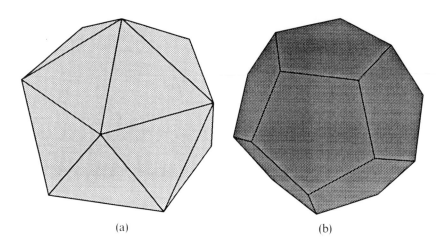

(a)   (b)

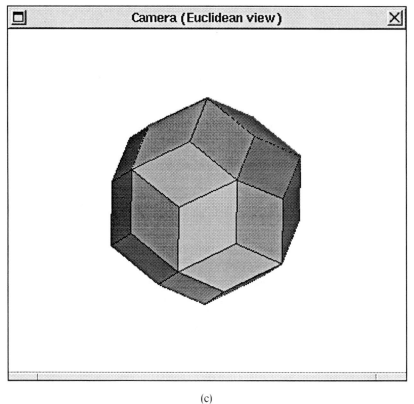

(c)

Fig. 1.2 (a) The regular icosahedron; (b) the regular dodecahedron. (c) The convex hull of the union of their vertices is a rhombic triacontahedron.

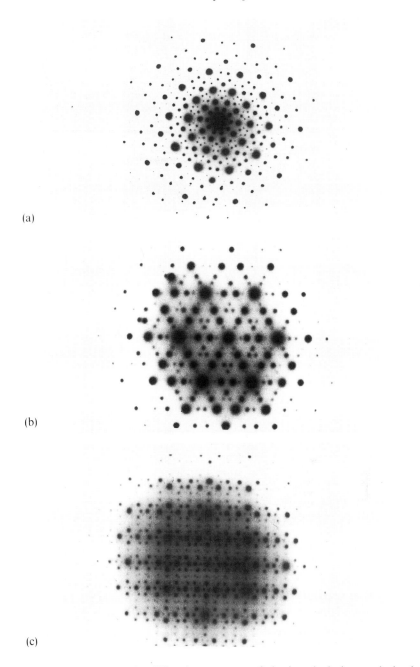

Fig. 1.3 Even very early diffraction patterns of the icosahedral crystal clearly showed (a) ten-fold, (b) six-fold, and (c) two-fold rotational symmetry. (Photographs courtesy of John Cahn.)

Fig. 1.4 The pyrite crystal is not a regular dodecahedron: notice that its faces are not regular pentagons.

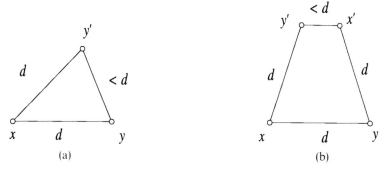

Fig. 1.5 (a) The angle of rotation cannot be less than $\pi/3$. (b) The angle of rotation cannot be $2\pi/5$.

  (ii) for any two regions R_1 and R_2 there is a translation that carries one to the other and the entire pattern onto itself;

  (iii) no two points interior to a region are translation-equivalent.

Periodicity is incompatible with five-fold rotational symmetry (and also with rotational symmetry of any order greater than six). To see why, let us assume that there exists a plane pattern that is both periodic and invariant under N-fold rotation. Let x be a rotation center. Since the pattern is periodic, it has a countable infinity of such centers and there is a a minimum distance d between them. Assume that y is another such center and that the distance between x and y is d : $|x - y| = d$. A counterclockwise rotation through $2\pi/N$ radians about x carries y to another rotation center y' and we must have $|y - y'| \geq d$. This is possible only if $N \leq 6$ (Figure 1.5(a)). If $N = 5$, then $|y - y'| > d$, but another problem arises. Since y is also a rotation center, clockwise rotation about this point must carry x to another center x'. But $|x' - y'| < d$ (Figure 1.5(b)).

The same argument is valid in three-dimensional space, since every rotation is a rotation of a plane about an axis orthogonal to it. We have thus established

Theorem 1.1 (the crystallographic restriction) Rotational symmetries of order greater than six, and also five-fold rotational symmetry, are impossible for a periodic pattern in the plane or in three-dimensional space.

The discovery of quasicrystals shattered this fundamental 'law', not by showing it to be logically false but by showing that periodicity is *not* synonymous with long-range order, if by 'long-range order' we mean whatever order is necessary for a crystal to produce a diffraction pattern with sharp bright spots. It suggested that we may not know what 'long-range order' means, nor what a 'crystal' is, nor how 'symmetry' should be defined. Since 1984, solid state science has been under going a veritable Kuhnian revolution (Kuhn, 1970).

But mathematical crystallography has been enriched by successive conceptual revolutions throughout its history. For example, the quotation from Barlow and Miers at the beginning of this chapter, taken from a report to the British Association for the Advancement of Science, reflects the authors' pride in a recent milestone, the enumeration of the three-dimensional crystallographic groups. That research, too, had been driven by the discovery that the concept of homogeneity had to be broadened in order to explain some then-newly discovered morphological properties of crystals.

As we continue to widen the scope of our inquiry and enlarge our definition of homogeneity, we confront again the questions that fascinated – and challenged –our predecessors. In this chapter we review some of this long and interesting history, and take tentative steps into a long and interesting future.

1.2 Ancient views on the structure of crystals

Before the seventeenth century, there was no science of crystals. Through the sixteenth century, crystals were prized mainly for their beauty and for their supposed medicinal and magical properties, as they had been for centuries (and are again in this 'New Age'). Sixteenth-century authors dwelt on the 'virtues of gems', citing authorities on these matters from the

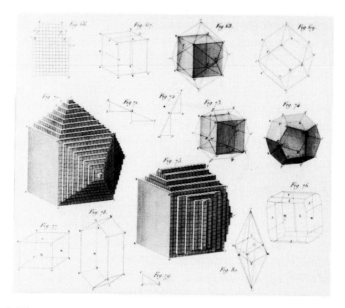

Fig. 1.6 'The regular forms of crystals suggest that building units are regularly arranged.' (This illustration is taken from Haüy (1822).)

ancients to their contemporaries; some of the details were quite bizarre (Metzger, 1922).

Today we are trained to think (Figure 1.6) that 'The regular forms of crystals suggest that within a crystal atomic building units, congruent to each other, are regularly arranged.' (Engel, 1986) Thus it is sometimes a surprise to learn that the regular forms of crystals seem not to have suggested any such thing to our predecessors. They may not have even noticed the regularity of the forms. We look for geometrical order; our early predecessors looked for other things. Aside from medicine and magic, classification does not seem to have been particularly important to them. When early writers did classify stones, they did so in ways that now seem strange to us, for example according to the plant or animal that they most resembled.

It is in another strand of the history of science – atomism – that we find ideas from ancient times that are compatible with present-day thinking about crystals. The problem of whether matter is discrete or continuous has been a subject of major importance to thinkers at least since the time of the ancient Greek philosophers. Early atomists grappled with such questions as the sizes of the ultimate particles, whether there were finitely

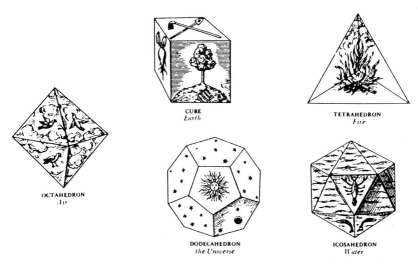

Fig. 1.7 Plato's assignment of the regular solids to the shapes of the four elements (and to the cosmos). (Adapted from a drawing by Johann Kepler.)

many or infinitely many particle prototypes, what they looked like, and how they could be joined together. The first detailed 'atomic' theory was that of Plato, who (in the *Timaeus*) ascribed the forms of four of the regular polyhedra to the particles of the four 'elements' (earth, air, fire, and water) (Figure 1.7). These choices were not entirely fanciful. Plato realized that his theory had to be able to explain phase transitions (such as ice to water to steam); to do this he hypothesized that his particles could be disassembled and their triangular facets reassembled into the other shapes. He also tried to match the shape to the properties of the element – thus water, which flows easily, should be made up of icosahedral particles, the most sphere-like of the five. Some aspects of modern theories of liquids and glasses are not so very different.

Plato's was the first 'application' of one of the greatest achievements of Greek mathematics, the discovery of the five regular convex solids. As Waterhouse (1972) has pointed out, *the* discovery was not that each of these shapes was found, but the recognition that together they form a special class. The Greeks may even have understood the duality of the cube and octahedron, and of the icosahedron and dodecahedron.

Whatever their motivation for this work may have been, the Greeks evidently considered the regular solids to be very important. These shapes were immortalized, in very different ways, by Plato and by Euclid, and this helped to preserve them in the poetic, artistic, and scientific consciousness of the Western world. They have functioned ever since as the Sirens in Ulysses, luring many explorers – not only Plato and, much later, Kepler, but also more modern ones – off course and into danger.

Aristotle, who was not an atomist, and who as a naturalist fully appreciated the diversity of nature, raised many objections to the details of Plato's theory as well as to its underlying premises. He argued, in *De Caelo*, that if indeed the supposed particles had the shapes that Plato assigned to them, then copies of each of the regular polyhedra should fill the space around a point (that is, they would be space-fillers at least locally). But, said Aristotle, since of the regular solids only the cube and the tetrahedron have this property, the theory must be false. Of course, the regular tetrahedron does not fill space; the efforts of Aristotle's followers to justify this error is a comical but instructive chapter in the history of mathematics (Struik, 1925; Senechal 1981). Was Aristotle's error due to the same arrogance toward hands-on geometry that still persists today? If he had bothered to build a model he would have seen his error immediately (Figure 1.8).

But even though Aristotle got the details wrong, the point he was trying to make was right: if we assume that space is completely filled by particles, then their shapes must be such that the particles can indeed fill it. Very few kinds of three-dimensional shapes are candidates (Grünbaum and Shephard, 1980). The importance of this question was not recognized until the nineteenth century. Indeed, both because of Aristotle's refutation and the broader rejection of atomism, there seems not to have been any serious thinking about the geometrical structure of matter until the seventeenth century.

The first treatise on crystals that is modern in any sense of the word was a short work by Kepler on the structure of snowflakes, written in 1611 and presented as a New Year's gift to a friend. In this work, Kepler tried to explain why snowflakes are hexagonal (Kepler, 1611). To do this, he

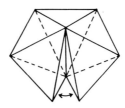

Fig. 1.8 Contrary to Aristotle's contention, five regular tetrahedra cannot be fitted together around an edge.

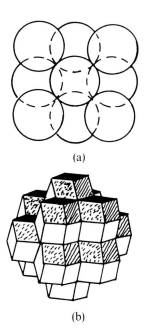

(a)

(b)

Fig. 1.9 (a) Spheres in cubic closest packing. (b) When uniformly compressed, the spheres become rhombic dodecahedra.

assumed that they are composed of tiny spheres (whether these were supposed to be atoms or other entities he did not say). Kepler recognized the analogy with the honeycombs of bees, whose cylindrical wax cells acquire planar facets when packed closely together. By compressing spheres arranged in what we now call cubic close packing, Kepler found that as they flattened they turned into rhombic dodecahedra. Thus rhombic dodecahedra fill space (Figure 1.9).

Kepler's views were highly unusual for his time: none of the concepts basic to our contemporary notion of a crystal was widely accepted then. In those days, the word 'crystal' still referred only to quartz, which was thought to be permanently frozen ice. 'Crystallography' as a generic term was not proposed until 100 years later. Today it is clear to us that a crystal is different from a fossil, that a crystal is a relatively simple, inanimate, modular structure, and that the modules of which it is composed are not fictitious, but are really there. Four hundred years ago, not only were crystals not distinguished from fossils, they were not always distinguished from living things. The modular concept of crystals did not gain acceptance until 200 years after Kepler, and the reality of the modules (i.e. atoms) was not fully established until the discovery of x-ray diffraction in 1912.

We can gain some understanding of competing views of that era by studying one of the most interesting seventeenth-century treatises on crystals, the *Prodromus* of Nicolaus Steno (1669). This is a brief work 'concerning a solid body enclosed by the process of nature within a solid'; it contains Steno's views on crystals, fossils, and other geological matters. Steno wanted to explain how crystals came to be inside a fissure or rock, and this led him to speculate on crystal growth. His views on the subject seem strikingly modern. For him, matter was particulate and crystal growth was modular. A crystal grows in solution; particles in the solution are attracted to the crystal and migrate around on the surface until they find suitable places to settle down. Some surfaces grow faster than others, and although they appear smooth to our eyes, on the scale of the particles they are rough.

Although Steno talked about experiments, he does not seem to have performed any. Instead, his main purpose seems to have been to create a coherent theory, one which, he realized, would be novel for that time. And he made this very clear: 'No room at all is left here', he wrote, 'for the belief of those who affirm that crystals grow plant-like, by nourishment.'

Metzger (1922) has summarized beautifully the many theories of vegetative growth of minerals of those days. In the absence of experiment, arguments by analogy were given free rein. As late as 1702 Tournefort, one of France's leading botanists, could write: 'It seems that we cannot doubt that some stones have internal organization; they draw liquid nourishment from the earth; this juice must be carried from their surface, which should be regarded as a kind of skin, to all the other parts ... there is no more difficulty in imagining this than in understanding how the sap passes from the roots of our largest oaks and pines to the tips of their highest branches.'

Silly as this may sound, even today we still have only a vague notion of the concept and meaning of 'organization'. (It is sometimes said that organization is the characteristic of life. Some seventeenth-century natural philosophers may have thought the same thing, and then completed the syllogism: crystals are organized, therefore crystals are alive, or at least some organic agent is responsible for crystallization and growth. This may explain their classification by resemblances to living things.)

Such was the force of these arguments that 100 years after Steno, the famous botanist Linnaeus (1768) paid at least lip service to them: in his classification of stones he declared salts to be their fathers and earths to be their mothers! Even if he was speaking metaphorically, his choice of metaphor is very revealing.

This digression into now-discarded ideas shows us that what we now call 'defects' in crystal structure, such as the 'veins' one can sometimes see inside a translucent crystal, were at one time thought to be key features and thus much more important than symmetry or shape. Only at the end of the eighteenth century, with the invention of the contact goniometer and the discovery of important commonalities in the myriad of forms that can be assumed by crystals of the same substance, did attention shift to crystal shape and then, as we will see, to symmetry. (While everyone agreed on the importance of symmetry, there was no agreement on modularity. Emphasis on external characteristics was only gradually subordinated to regularity of structure on the atomic level; regularity of internal structure was consolidated as the ruling paradigm only after the discovery of x-ray diffraction.)

Interestingly, the idea that defects might be more important than regularity has been raised in our own time, in order to explain features of crystal growth that otherwise seem anomalous; defects are also essential for understanding mechanical properties. Defects in quasicrystal structures continue to receive serious attention.

1.3 From building-blocks to lattices

Several years after his musings on the snowflake, Kepler again studied packing problems, this time the packing of polygons in the plane (Kepler, 1619). In the course of his investigations (Figure 1.10), he discovered – among other things – the 11 Archimedean tilings of the plane (tilings by regular convex polygons, not necessarily all alike, but with identical arrangements at each vertex). Since deepest antiquity, the designers of mosaic patterns have known that the plane can be paved with squares, equilateral triangles, and regular hexagons, and some combinations of these with other polygons, but as far as we know Kepler was the first to study the question in a systematic way. Evidently he was not motivated by crystal problems this time, but his drawings are now considered part of crystallography because he was obviously searching for regularity. The Archimedean tilings were rediscovered by crystallographers in this century. But it is tiling Aa that seems especially remarkable today. It is closely related to the Penrose tilings of the plane (Chapter 6).

Half a century after Kepler, in England, Robert Hooke (1665) drew pictures of circles packed in various polygonal shapes to illustrate his ideas

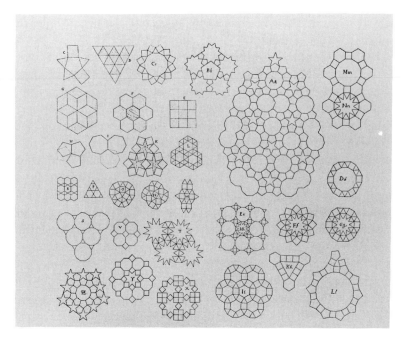

Fig. 1.10 Kepler's experiments with polygon packings. The Archimedean tilings
are labeled D, F, E, L, P, N, M, S, V, Ii, and Mm.

on the structure of crystals (Figure 1.11). Hooke struck the keynote theme
for crystallography up to our own time with his casual remark that

Had I time and opportunity I could make probable, that all these regular Figures
that are so conspicuously various and curious, and do so adorn and beautifie such
multitudes of bodies, arise onely from the most plain, obvious and necessary
conjunctions of such figur'd particles that are possible.

Hooke's drawings were similar to the drawings of several of his near
contemporaries, most notably Christiaan Huyghens (1690). Huyghens,
intrigued by the double refraction of light by calcite crystals, showed how
the strange behavior of light in this crystalline medium could be accounted
for if the crystals were built of ellipsoidal particles. None of these
hypotheses, or others like them, led directly to any serious investigations
of packings. The drawings were suggestive however, and undoubtedly
influenced later thinkers. For example, Steno is sometimes credited with
the discovery of the 'law of constancy of interfacial angles' (Figure 1.12),
although his observations were only noted in the captions to some of his

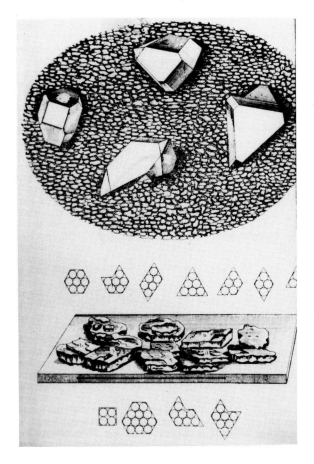

Fig. 1.11 Robert Hooke shows how different crystal forms can result from the same type of sphere packing. (From *Micrographia*, 1665.)

drawings in the *Prodromus*. The law, which states that despite variations in external form, the angles between the faces of the crystals of a given substance are invariant, was formulated more generally a century later, after the invention of the goniometer.

This discovery of regularity underlying apparent variation led directly to the building-block theory of crystal structure. At the turn of the nineteenth century, R. J. Haüy (1822) proposed the hypothesis that each crystal is composed of building-blocks of a characteristic polyhedral shape which are stacked together in a regular way to build the crystal shape (Figure 1.6). Haüy was able to explain, for the first time and to an approximation that was good in some cases (though not so good in

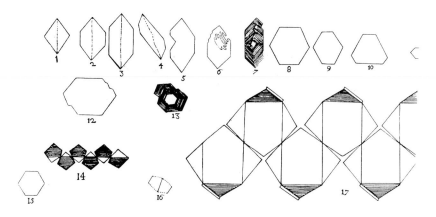

Fig. 1.12 Steno's drawings of iron crystals 'anticipate' 'the law of constancy of interfacial angles'.

others), why real crystals have the forms they do. In particular, he showed that the faces and vertices of 'building-block crystals' can always be assigned rational coordinates in some coordinate system. Since the regular icosahedron and dodecahedron cannot be assigned rational coordinates in any system, building-block crystals cannot have this symmetry.

When it was first introduced, Haüy's theory met with the same reception as Plato's; it was vigorously attacked by his contemporaries and even by some of his students. But their objections were not the ones suggested today by the discovery of aperiodic crystals. One problem for Haüy's contemporaries was his implicit atomism, which displeased German scientists influenced by Naturphilosophie. Other critics accepted the hypothesis of discrete crystal structure but objected to the details of Haüy's blocks. They argued that one need not and should not postulate blocks of a definite size and shape; it is only necessary to assume that solid matter is built of discrete particles arranged in rows and layers. These particles can be represented by points located at their centers of gravity. This is the viewpoint that quickly prevailed; periodicity was soon the cornerstone of crystal theory. We will call it the *lattice hypothesis*.

1.4 The ascendance of symmetry

The argument given above to show that five-fold symmetry is impossible in a periodic pattern did not really invoke periodicity *per se*. We only needed to assume that the point set under consideration is discrete, that is, that there is a minimum distance between points in the set under consideration; then, by studying the pattern of points generated by two rotations, we were able to contradict this assumption. In the nineteenth century, discrete point sets generated by 'rigid' motions were called *regular systems of points*. After Haüy, the regular system of points became the central concept of theoretical crystallography, and their classification was a primary problem. The scope of these investigations was gradually broadened, together with the concept of rigid motion.

The simplest regular system of points is the point lattice $\mathcal{L}^p$. Point lattices are characterized by a special property: all points are translation-equivalent. That is, any point of $\mathcal{L}^p$ can be carried to any other by a translation that at the same time maps $\mathcal{L}^p$ to itself. The 'natural' classification principle for lattices is symmetry (see Michel and Mozryzmas, 1989, for a contemporary account); Figure 1.13 shows the symmetry types of two-dimensional point lattices.

In (1850) Bravais showed that there are exactly 14 symmetry classes of point lattices in three-dimensional space and pointed out many crystallographic applications of the lattice hypothesis (Bravais, 1851), including a prediction of the faces most likely to appear on the macroscopic crystal.

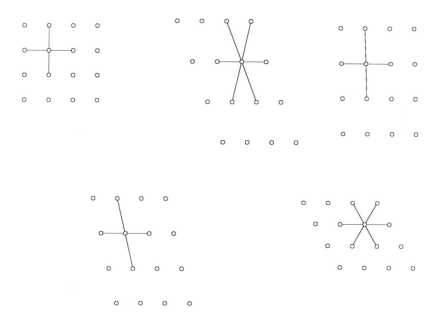

Fig. 1.13 The symmetry types of two-dimensional point lattices.

These applications of course helped to make the lattice hypothesis plausible and acceptable to crystallographers. Bravais brought elegance to theoretical crystallography by blending geometry and algebra in a way that we now call group action (Appendix I): he studied the ways in which rigid motions permute points on a sphere or in space. In contemporary language, Bravais thought of the point sets as orbits of symmetry groups.

Like others past and present who have made fundamental contributions to this subject, Bravais was neither a professional mathematician, nor was he a professional crystallographer. Indeed, it would be as difficult to classify him as it is to classify crystals today: he was a naval officer, botanist, meteorologist, and explorer, among other vocations. His place in the hagiography of crystallography is secure, but his role in the development of group theory – and that of symmetry more generally – is rarely mentioned in histories of mathematics, although the development of symmetry concepts was stimulated by the symbiotic interaction of Bravais and the Parisian mathematicians of his time. Several of Bravais' papers were presented to the Academy by A. L. Cauchy and, more importantly, they influenced the work of some of its members. In particular, Bravais's memoir on lattices came to the attention of Camille Jordan, who used it as the starting point of his *Mémoire sur les groupes des mouvements* (Jordan, 1868).

Jordan realized, as did contemporary crystallographers, that it was necessary to enlarge the definition of a regular system of points from the lattice to any point set whose points were equivalent under translations *or* rotations. He also moved the focus of attention from the point set to the motions that generate it, that is, from the orbit to the group itself.

Jordan studied the groups of motions (rotations, translations, and screw rotations) that carry regular systems onto themselves. He did not enumerate the groups nor did he discuss applications of his work to crystallography, although he was aware of them; as he pointed out, the groups could be used 'to form in all possible manners the systems of self-superposable molecules in different positions'. Crystallographers realized that Jordan's paper was just what they needed.

By 1891, after a tortuous history, the concept of a three-dimensional regular system of points had been extended to point sets whose symmetries included not only the motions studied by Jordan but also reflections, glide reflections, and rotary reflections, and the 230 generating groups – now known as the crystallographic groups of E^3 – had been enumerated, for the most part independently, by Fedorov and Schoenflies (Fedorov, 1891; Schoenflies, 1891).

These two very different men approached the enumeration in very different ways. Schoenflies, who was originally directed to the problem by Felix Klein, saw it as an exercise in group theory and indeed as a nice example of the power of Klein's Erlangen Programm. This 'program', announced by Klein in 1872, was in fact a declaration that geometrical objects should be classified by the features that are left invariant when transformations are performed on them. From this point of view, geometry is subsumed under group theory. The crystallographic groups were an attractive example precisely because they could be studied without having to engage in the ferocious nineteenth-century debates over just what the building-blocks of real crystals were or what shapes they had. (As Schoenflies put it, 'Within the fundamental domain the crystallographer can do as he likes.')

The impact of Klein's pronouncement was felt throughout all of mathematics; it has governed the study of geometry for the past century and is only now being challenged (on several fronts). The emphasis on groups may well have retarded the development of some areas of geometry by focussing attention on symmetry rather than shape, on invariants rather than special properties. For example, group theory cannot answer a question that seems fundamental today: which shapes tile space and in what ways?

Fedorov, on the other hand, had a low opinion of symmetry as a classification principle because – in today's language – it is not generic. Although he devoted much time to deriving the crystallographic groups (using his own idiosyncratic system of analytic geometry), this work was much less important to him (and is much less important to us) than his study of space-filling polygons and polyhedra.

Like Bravais, Fedorov was neither a mathematician nor (at first) a crystallographer, but like Bravais he was well aware of recent mathematical developments and used them to make profound contributions to mathematical crystallography. Perhaps the most important of his many works was a book that he began to write as a boy of 16, *An Introduction to the Theory of Figures* (Fedorov, 1885; Senechal and Galiulin, 1984).

The most important section of the book is the fourth, in which Fedorov took up the problem, first raised implicitly by Aristotle and central to Haüy's building-block theory, of determining the properties of polyhedra that fill space by translation. Fedorov was a much more sophisticated mathematician than Haüy, but he shared Haüy's belief in polyhedral building-blocks. Assuming that a crystal consists of discrete identical units ('crystal molecules') located at the points of the crystal lattice, Fedorov showed that the space that each unit 'occupies' is a convex polyhedral region with the lattice point as its center. These regions, which he called parallelohedra, are congruent and fit together along whole faces to fill all of space. In his view, the first task of the structural crystallographer was to determine the parallelohedron of each crystal. Using equations derived from Euler's fundamental formula relating the numbers of faces, edges, and vertices of a convex polyhedron, together with a variety of geometrical techniques, Fedorov found that there are two combinatorial types of parallelogons in the plane (the quadrilateral and the hexagon), and exactly five types of parallelohedra in three-dimensional space – the cube, the rhombic dodecahedron, the elongated rhombic dodecahedron, the 14-face truncated octahedron,and the hexagonal prism (Figure 1.14).

Fedorov devoted much of his life to the building-block problem; to the end he believed that his greatest work was *Das Kristallreich* (Fedorov, 1920), a huge treatise in which he applied the theory of parallelohedra to much of the mineral kingdom. However, his masterpiece was obsolete before its (posthumous) publication. With the discovery of x-ray diffraction in 1912, the crystallographers' interest in deducing crystal structure from crystal form was abandoned overnight, and with it went a great deal of the motivation for studying tilings.

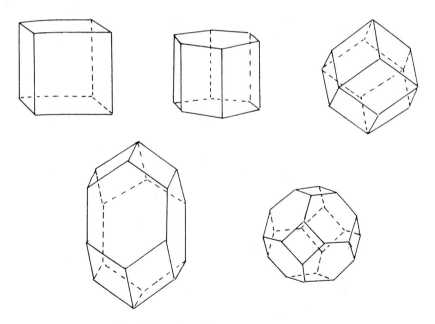

Fig. 1.14 Fedorov's five parallelohedra.

1.5 Hilbert's 18th problem

In 1900, the eminent mathematician David Hilbert presented a list of 23 problems to the International Mathematical Congress in Paris which he hoped would guide mathematical research in the twentieth century. One of the most interesting (though perhaps the least celebrated) of these problems was the 18th, which has its roots both in nineteenth-century crystallography and in the work of Hilbert's friend Minkowski. Although he does not appear to have been particularly interested in crystals, it was the work of Minkowski, more than that of any of his contemporaries, that contributed to the further development of mathematical crystallography. More than any specific result, his study of n-dimensional lattices in connection with the geometry of numbers created a mathematical atmosphere receptive to questions about lattices and space-filling polytopes.

Minkowski was evidently unaware of Fedorov's work on space-filling polyhedra. He too investigated parallelohedra, not in order to enumerate them but, rather, to determine some of their basic properties. By a simple and beautiful convexity argument (1907), he established the fundamental result that the number of $(n-1)$-dimensional faces of a convex n-dimensional parallelotope is at most $2(2^n - 1)$. Something of a synthesis of the work of Fedorov and Minkowski was later achieved by G. Voronoï (1908;1909), who discovered many of the properties of n-dimensional parallelotopes (see Chapter 2).

By 1900, then, the three-dimensional crystallographic groups were known, the study of space-filling polyhedra had left the realms of wishful thinking and become a mathematical subject, and the generalization of some of these problems to n dimensions had begun. Hilbert did not contribute directly to any of them, but his role in the development of mathematical crystallography was none the less pivotal: by gathering these strands together and shaping them into a problem for his influential list he brought them to the attention of the mathematical community. Hilbert asked:

A. Is there in n-dimensional euclidean space ... only a finite number of essentially different kinds of groups of motions with a fundamental region?

B. Whether polyhedra also exist which do not appear as fundamental regions of groups of motions, by means of which nevertheless by a suitable juxtaposition of congruent copies a complete filling up of all space is possible.

C. How can one arrange most densely in space an infinite number of equal solids of given form ... that is, how can one so fit them together that the ratio of the filled to the unfilled space may be as great as possible?

The full text appears in Section 7.

Each of Hilbert's questions is relevant to the material of this book.

Question A was certainly prompted by the successful enumeration of the three-dimensional crystallographic groups; it asks whether there is a finite number of them in every dimension. (By a group with a 'fundamental region' Hilbert meant an infinite group of motions whose orbits are discrete.)

The question was answered in the affirmative by Bieberbach and Frobenius who showed, in a series of papers published between 1910 and 1912, that the number of isomorphism classes of such groups in E^n is finite for all positive integers n. More importantly, they discovered the structure of these groups and the geometric structure of their orbits, the regular systems of points. (See (Bieberbach, 1910, 1912) and (Frobenius, 1911).)

Bieberbach's fundamental result had already been established by Schoenflies for the special case $n = 3$:

Theorem 1.2 Every regular system of points in E^n is a union of a finite number of congruent lattices.

It follows that any group G of motions whose orbits are discrete has an invariant translation subgroup T, and the quotient group G/T is isomorphic to a finite group P of isometries that permute, each in a different way, the set of lattices. This result served to further entrench the lattice as *the* fundamental concept of crystallography.

Question B asks whether there exists a tile such that no symmetry group acts transitively on any tiling in which it appears. It has been solved, or remains unsolved, depending on how you interpret it.

All that Hilbert asked for was a tile that is not a fundamental region for any group. A three-dimensional example of such a tile was found by Reinhardt (1928); later, a two-dimensional example was found by Heesch (1935). Since then more examples have been found, some of them convex. In all cases, the tilings are periodic. The most charming example is surely Escher's 'Ghosts' (Figure 1.15). If you look closely you will see that this complicated-looking tiling is really just a modification of a tiling of the plane by equilateral triangles! Each triangle has two ghosts on each edge and three ghosts in its interior. Since an interior ghost cannot be mapped onto an edge ghost by any rigid motion, the symmetry group of this tiling does not act transitively on the tiles. Moreover, the ghost tiling is uniquely determined by its single prototile: Figure 1.15 shows the only way that the

Fig. 1.15 Escher's 'Ghosts'. (M.C. Escher Foundation – Baarn – Holland. All rights reserved.)

ghosts can be fitted together. Thus it is a solution to Question B, in the most literal interpretation of that question.

A more interesting and challenging variant of question B is: *is there a single tile which builds only nonperiodic tilings?* There is a three-dimensional tile that tiles space only nonperiodically; we will discuss it and its tilings in detail in Chapter 7. No two-dimensional example of a single tile with this property is known, but there are several *pairs* of tiles that tile only nonperiodically, including the famous Penrose tiles (see Chapters 6 and 7).

Question C is very broad. Indeed, since the densest possible packing of congruent bodies is a tiling, it asks, among other things, which shapes tile space. However, Question C is usually interpreted as the problem of finding dense packings of equal spheres. Sphere-packing problems are receiving increasing attention, partly because of their close relation to coding theory (Conway and Sloane, 1988). Recently, a proof was announced of the long-standing conjecture that the cubic closest packing is of maximal density in E^3; as of this writing, this claim is controversial. But whether or not the conjecture is true, many important questions about packings of unequal spheres in E^3, and packings of polyhedra, remain unsolved.

In addition to seeing the solution of part A of Hilbert's problem, the year 1912 was an important one for crystallography for another reason: Max von Laue showed that crystals could serve as diffraction gratings for x-rays. This monumental experiment settled three controversial questions simultaneously: it proved that x-rays are a form of light, that atoms exist, and that the atoms in crystals are arranged in an orderly way.

Diffraction photographs turned out to be the Rosetta stone of the solid state, because from the diffraction images one can deduce the symmetry and in some cases even the positions of atoms in the crystal structure. Now, in addition to algebra and geometry, crystallographers needed analytic methods – especially Fourier analysis – in order to explain diffraction patterns. Since crystal structures were assumed to be periodic, they could be modeled by periodic functions and hence by Fourier series. (A function of one variable, $f(x)$ is periodic with period ω, if

$$f(x + \omega) = f(x)$$

for all $x \in R$. Any periodic function can be expanded into a Fourier series:

$$f(x) = \sum_n c(n) \exp(2\pi inx), \tag{1.1}$$

where the nth Fourier coefficient is $c(n) = \int_n^{n+1} f(x) \exp(-2\pi inx)\, dx$. The generalization to several variables is relatively straightforward.)

Fourier transformation has turned out to be even more important; recognition of its role in the interpretation of x-ray diffraction patterns has helped to make x-ray diffraction one of the most powerful scientific tools of our time (see Chapter 3).

Another twentieth-century development, one that may not have appeared relevant to crystallography at the time it was introduced but which is seen as important today, is the theory of *almost periodic functions* (Bohr, 1924, 1925, 1933).

Instead of periods, an almost periodic function has 'almost periods': for example a function of one variable f is almost periodic if, for every $\epsilon > 0$, there is a relatively dense set of real numbers ω_ϵ (see Definition 1.4 below) such that

$$\sup_{x \in R} |f(x + \omega_\epsilon) - f(x)| \leq \epsilon. \tag{1.2}$$

Bohr proved that, like periodic functions, an almost periodic function can be expanded in a 'Fourier series': in this case the series has the form

$$f(x) = \sum c(n) \exp(2\pi i \lambda_n x), \tag{1.3}$$

where (λ_n) is a sequence of real numbers. Moreover, Bohr showed that almost periodic functions are closely related to periodic ones: every almost periodic function is the diagonal function $g(x, x, \ldots)$ of a 'limit periodic' function g of a finite or countable number of variables. (A limit periodic function is the uniform limit of a sequence of periodic functions.) The sums (1.3) arise naturally in the study of diffraction patterns of aperiodic crystals (see Chapter 3).

At about the same time, in the 1930s, the Russian mathematician B. N. Delone (also spelled Delaunay), together with a group of students and colleagues, undertook a reconstruction of the fundamental concepts of classical crystallography from first principles (Delone, Aleksandrov, and Padurov, 1934). Over the next 45 years they clarified and extended Fedorov's work on tilings, and developed a coherent body of *n*-dimensional mathematical crystallography, incorporating the work of Minkowski, Voronoï, and others.

According to Delone, the study of crystallography should begin with very general point sets, constrained by only two simple, physically reasonable properties, 'discreteness' and 'homogeneity'. No regularity or other symmetry conditions are imposed. We will call such sets *Delone sets* (see also Definition 1.5 below).

Discreteness is a 'hardcore condition': it expresses the fact that the centers of the atoms in any gas, liquid, or solid, cannot come arbitrarily close together. The 'homogeneity condition' models the fact that atoms in these phases tend to distribute themselves more or less uniformly

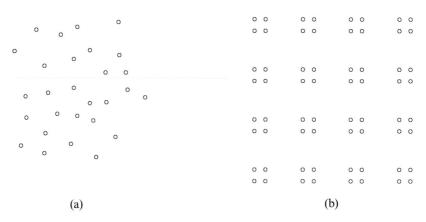

(a)   (b)

Fig. 1.16 Portions of two Delone sets: (a) amorphous, (b) symmetrical.

throughout the available space. Discrete, homogeneous point sets can serve as models for a broad range of structures, from highly amorphous to highly symmetrical (Figure 1.16).

The Delone school made many contributions to tiling theory and other crystallography-related areas of geometry. One of its most intriguing discoveries (Delone, Dolbilin, Shtogrin, and Galiulin, 1976) was that regular systems of points are determined by local configurations: if the spherical neighborhoods about each of the points of a Delone set are congruent up to a certain finite radius, then the infinite neighborhoods about the points are also congruent, which means that there are motions that carry any point to any other and the whole point set onto itself. Thus this theorem says that 'global' regularity results from the structure of 'local' configurations. Of course, the exact radius required for a given set depends on its specific configuration. Delone also generalized some of Fedorov's results to four dimensional space.

As Bieberbach's theorem (Theorem 1.1) shows, theoretical crystallography can be conducted in n-dimensional space for any n. But it was not until the 1970s that anyone thought of applying n-dimensional crystallography to the study of three-dimensional crystals (see Section 1). Before the end of that decade, a table of all 4895 crystallographic groups of four-dimensional space was published (Brown *et al.*, 1978) in response to the need of crystallographers studying modulated structures.

1.6 Toward a new definition of 'crystal'

Looking back over all this history, we see that ideas about crystal-geometry have never been static. Vegetative theories aside, we can say that a major theme has been a quest to relate the regularity of images (form, diffraction pattern) to regularities of structure (sphere packings, arrangements of building-blocks).

In that spirit, it is time to put the definition of 'crystal' that has been with us since the early nineteenth century into the historical archives. But what should take its place?

The International Union of Crystallography, an august body that encourages and keeps tabs on crystallographic research all over the world, has established a Commission on Aperiodic Crystals to facilitate communication among, and set publication standards for, those engaged in experimental and theoretical research on aperiodic crystals. The working definition of 'crystal' proposed by this commission in 1992 is:

Definition 1.1 (informal) A crystal is any solid with an essentially discrete diffraction diagram.

Deliberately, the commission did not build symmetry or regularity of any kind into the new definition of 'crystal'; it only required, implicitly, that the structure have whatever characteristics are necessary and sufficient to produce a diffraction pattern with sharp spots. Symmetry is not abandoned; but in keeping with the new definition the focus is shifted from structure to image. Thus

Definition 1.2 The symmetry of a crystal is the symmetry implied by its diffraction diagrams.

Like any good generalization, the above definitions include periodic crystals as special cases. In a series of papers, Mermin and his colleagues have generalized the regular systems of points to include the orbits of Z-modules (see Chapter 2); from this they show how one can construct all the groups needed for a generalized crystallography without explicit recourse to higher-dimensional spaces (see e.g. (1991; 1992); further references can be found in these papers).

It is the task of mathematical crystallography to take up where Definitions 1.1 and 1.2 leave off and to determine what they imply for crystal structure. In the light of Definition 1.1, the 'crystals' we consider will be Delone sets with enough local uniformity to produce sharp spots in diffraction patterns.

This suggests that we must study the properties of local configurations. First we need a formal definition of a discrete point set.

Definition 1.3 A point set $\Lambda \subset E^n$ is said to be discrete if there exists a positive real number r_0 such that, for every $x, y \in \Lambda$, $|x - y| \geq 2r_0$.

It follows immediately from the definition that

Proposition 1.1 If Λ is discrete, then it has the local finiteness property: for every closed ball $\overline{B}_x(r)$ with center $x \in E^n$ and radius r, $\overline{B}_x(r) \cap \Lambda$ is a finite set.

The converse of this proposition is false, since a set with the local finiteness property may have an accumulation point 'at infinity'.

Delone's 'homogeneity' condition is also known as relative density.

Definition 1.4 A point set Λ is relatively dense in E^n if there is a positive real number R_0 such that every sphere of radius greater than R_0 contains at least one point of Λ in its interior.

Definition 1.5 A point set $\Lambda \subset E^n$ is a Delone set if it is discrete and relatively dense.

The parameter R_0 is called the covering radius of Λ.

Delone sets were called (R, r) systems by Delone and his colleagues.

Proposition 1.2 A Delone set is countably infinite.

Proof We can tile E^n by a countable number of congruent cubes. Since Λ is discrete, each cube can contain only a finite number of points of Λ. Thus Λ itself is countable. Relative density ensures that Λ is infinite. ❑

The local configurations in Delone sets can be characterized in various ways. Usually we will consider 'circular' configurations centered at points of Λ:

Definition 1.6 Let $r > 0$. The *r*-star at $x \in \Lambda$ is the finite point set $\overline{B}_x(r) \cap \Lambda$.

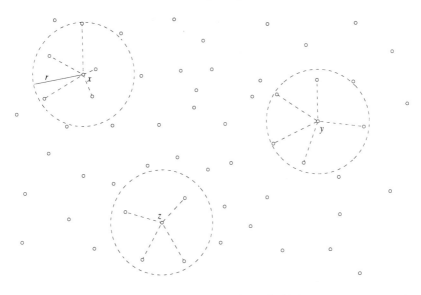

Fig. 1.17 *r*-stars for some points of a Delone set.

Figure 1.17 shows the *r*-stars for some points of a Delone set, for a particular value of *r*.

We need a more formal definition of a regular system of points somewhere in this book; this place seems about right.

Definition 1.7 A Delone set Λ is a regular system of points if the *r*-stars of its points are congruent for every $r > 0$ (or, equivalently, if Λ is the orbit of a crystallographic group).

Using this terminology, Delone's 'local criterion for regularity' says:

Theorem 1.3 There exists a critical radius $r_c > 0$ such that if the r_c-stars of a Delone set $\Lambda \subset E^n$ are congruent, then its *r*-stars are congruent for all $r > 0$. Thus Λ is a regular system of points.

It is often useful to think of the *r*-stars as copies of a basic set of *r*-'protostars', and to work with the set of protostars. We are free to decide what we mean by 'copy' in each context; usually one copies a star by translating it or by performing some other congruence. We assume that there is some appropriate group of motions $\mathcal{M}$ for making the copies.

Definition 1.8 The set of all *r*-stars of Λ, up to $\mathcal{M}$-equivalence, is called its *r*-atlas.

Since aperiodic crystals cannot be modeled by regular systems of points, we conclude that their *r*-atlases must contain more than one star, except possibly for 'subcritically small' values of *r*.

It should be possible to characterize crystals in terms of their *r*-atlases, that is, to determine what restrictions must be placed on the atlases if Λ is to produce a sharp diffraction pattern. However, this problem remains wide open. Some restrictions that have been suggested are:

• The atlases must be finite for all finite values of *r*. (It turns out, however, that this is neither necessary nor sufficient.)

• Λ should be repetitive. By repetitive we mean:

Definition 1.9 A Delone set is repetitive if, for all $r > 0$, the elements of the *r*-atlas are relatively dense in Λ.

Almost all of the nonperiodic Delone sets we will encounter are repetitive. It is a useful concept that can also be extended easily to tilings, packings, and other discrete structures. But there are repetitive patterns that do not produce sharp diffraction patterns.

• We might require self-assembly, since a theory of crystal structure should be able to give a plausible account of how the structure managed to assemble itself. The stars of an *r*-atlas might be thought of as interlocking pieces; by linking them together we build the crystal. But very little is known today about the formation of real aperiodic crystals; it is even possible that they do not 'assemble themselves' but, rather, are produced by phase transformations or other means (see Chapter 7). It seems premature to incorporate any growth mechanism in a working definition of crystal (although we will pay considerable attention to matching rules for tilings).

So, for the time being, we will not place these or any other restrictions on Λ. Instead, we assume that *a 'crystal' is a Delone set* Λ *with sufficient order to produce a diffraction pattern with bright spots, however that structure may eventually be characterized.* In Chapter 3 we will formulate the 'diffraction condition' – and thus the definition of 'crystal' – more precisely. This book is about point sets and tilings that do or do not satisfy this condition; the ultimate goal is to understand how to tell them apart.

We conclude with some useful terminology.

Definition 1.10 A *n*-dimensional **periodic crystal** is an *n*-dimensional regular system of points; it admits translations in *n* independent directions. An *n*-dimensional *aperiodic crystal* is one with no translational symmetry. A set that is periodic in *k* independent directions, $0 < k < n$, and aperiodic in the other $n - k$ is called *subperiodic.*

Quasicrystals are a special class of aperiodic crystals:

Definition 1.11 An *n*-dimensional quasicrystal is a crystal whose diffraction pattern exhibits symmetry forbidden by the crystallographic restriction for E^n (higher-dimensional analogues of Theorem 1.1 are discussed in Chapter 2.)

According to this definition, the quasicrystal phenomenon begins in E^2: there are no noncrystallographic rotational symmetries of the line.

1.7 Notes

1. Kuhn's *The Structure of Scientific Revolutions* has increased the self-awareness of scientists in every field. Although, like an impressionist painting, his portrait of scientific change dissolves on close inspection, it is widely regarded as a valid and important perspective. In particular, the use of the word 'paradigm' for the constellation of prevailing views of a scientific community at a given time has become commonplace.

2. In addition to Metzger's excellent book, an account of the history of geometric crystallography can be found in Senechal (1990).

For nearly two millenia Aristotle's commentators and followers tried to account for his nonsensical statement about tetrahedra without admitting that the Master could have been wrong. The question was finally resolved during the Renaissance.

The contribution of crystallography to the development of the theory of symmetry in the nineteenth century is rarely mentioned in histories of mathematics; a notable exception is Scholtz (1989).

Fedorov's book has never been translated from the Russian; indeed Fedorov had great difficulty publishing it in his own country (it is a very difficult book to read: his style is difficult to follow, and his proofs are confusing and not always correct). Still it is not difficult to recognize it as a work of genius, especially when one realizes that while he was working on it he was still a student (of medicine, physics, and chemistry) and then an army officer.

3. The full text of Hilbert's 18th problem, in English translation, is:

If we enquire for those groups of motions in the plane for which a fundamental region exists, we obtain various answers, according as the plane considered is Riemann's, (elliptic), Euclid's, or Lobachevsky's (hyperbolic). In the case of the elliptic plane there is a finite number of essentially different kinds of fundamental regions, and a finite number of congruent regions suffices for a complete covering of the whole plane; the group consists indeed of a finite number of motions only. In the case of the hyperbolic plane there is an infinite number of essentially different kinds of fundamental regions, namely, the well-known Poincaré polygons. For the complete covering of the plane an infinite number of congruent regions is necessary. The case of Euclid's plane stands between these; for in this case there is only a finite number of essentially different kinds of groups of motions with fundamental regions, but for a complete covering of the whole plane an infinite number of congruent regions is necessary.

Exactly the corresponding facts are found in space of three dimensions. The fact of the finiteness of the groups of motions in elliptic space is an immediate consequence of a fundamental theorem of C. Jordan, whereby the number of essentially different kinds of finite groups of linear substitutions in n variables does not surpass a certain finite limit dependent upon n. The groups of motions with fundamental regions in hyperbolic space have been investigated by Fricke and Klein in the lectures on the theory of automorphic functions, and finally Fedorov, Schoenflies, and lately Rohn have given the proof that there are, in Euclidean space, only a finite number of essentially different kinds of groups of motions with a fundamental region. Now while the results and methods of proof applicable to elliptic and hyperbolic space hold directly for n-dimensional space also, the generalization of the theorem for Euclidean space seems to offer decided difficulties. The investigation of the following question is therefore desirable: Is there in n-dimensional Euclidean space also only a finite number of essentially different kinds of groups of motions with a fundamental region?

A fundamental region of each group of motions, together with the congruent regions arising from the group, evidently fills up space completely. The question arises: Whether polyhedra also exist which do not appear as fundamental regions of groups of motions, by means of which nevertheless by a suitable juxtaposition of congruent copies a complete filling up of all space is possible. I point out the following question, related to the preceding one, and important to number theory and perhaps sometimes useful to physics and chemistry: How can one arrange most densely in space an infinite number of equal solids of given form, e.g., spheres with given radii or regular tetrahedra with given edges (or in prescribed positions), that is, how can one so fit them together that the ratio of the filled to the unfilled space may be as great as possible?

4. Fedorov and Schoenflies worked independently (and in ignorance of each other) for several years; when they learned that they were engaged in the same project they compared notes and corrected their mistakes (Burckhardt, 1971).

5. Our definition of 'repetitive' is due to Danzer (1991), but this term is far from standard: repetitive sets are also called 'quasiperiodic', 'almost periodic', 'recurrent', or 'set with the local isomorphism property'. This last phrase is also used for *families* of point sets (or tilings or packings): two point sets are said to belong to the same local isomorphism *class* if they are superimposable over arbitrarily large spheres. We will reserve this expression for such classes. (For a more formal definition of local isomorphism class, see Chapter 5.) The term 'species' is sometimes used for 'local isomorphism class', but we choose not to use it: although 'species' is briefer, it brings to mind seventeenth century notions of crystal biology!

6. The definition of crystal proposed by the Commission on Aperiodic Crystals, then *ad interim*, is contained in its 1992 document, Terms of Reference, published in *Acta Crystallographica* A, Vol. 48, 922–46. At that time the members of the commission were J. M. Perez-Mato (Spain), Chair; G. Chapuis (Switzerland), W. Steurer (Germany), M. Farkas-Jahnke (Hungary), and M. Senechal (USA).

7. Definition 1.11 is nonstandard – indeed there is no standard definition of 'quasicrystal'. Notice that the word can be interpreted in several ways. In one interpretation, 'quasi' means 'semi' or 'not quite'; it suggests that these new materials are not really crystals. Definition 1.1 rejects this interpretation while acknowledging that these solids are not periodic. Alternatively, 'quasi' may be short for 'quasiperiodic', which means, in this context, that its diffraction spots are linear combinations of a finite set of 'independent' points. The name 'quasicrystal' was introduced by Paul Steinhardt, with the second interpretation in mind.

8. Actually, the idea of studying crystals by embedding them in a higher dimensional space must be credited to Kepler who wrote, in 'The six-cornered snowflake' (1611),

While these starlets are falling, they consist of three feathered diameters, joined crosswise at one point, with their six extremities equally distributed in a sphere; consequently they fall on only three of the feathered prongs, and tower aloft with the remaining three, opposite those on which they fall, on the same diameters prolonged, until those, on which they rested, buckle, and the remainder, until then upright, sag onto the level with the former in the gaps between them.

2

Lattices, Voronoï cells, and quasicrystals

> A new problem, especially when it comes from the outer world
> of experience, is like a young twig, which thrives and bears fruit
> only when it is grafted carefully and in accordance with strict
> horticultural rules upon the old stem, the established achieve-
> ments of our mathematical science.
>
> *(David Hilbert)*

2.1 Introduction

Although the distinguishing characteristic of a nonperiodic crystal is the
absence of a translation lattice, we cannot dispense with this scaffolding
altogether. In the first place it serves as a point of departure for some of
the structures that we will consider, and secondly it reappears in the
emerging theory as a sort of higher-dimensional superstructure within
which the nonperiodicities appear in a natural way. So this chapter is
fundamental to the subject of the book even though it is possible that
eventually it will be seen as irrelevant to it.

Unless otherwise stated, we will work in n-dimensional Euclidean space
E^n with orthonormal basis $\vec{e}_1, \ldots, \vec{e}_n$. Points will be denoted by lower case
letters $(a, b, \ldots, x, y, z)$ and the corresponding vectors $\vec{a}, \vec{b}$ will always be
the vectors from $0 \in E^n$ to the point with the same letter. We will use the
terminology of group action (such as 'orbit') freely; it is explained in
Appendix I.

Recall that a set of k vectors in E^n is linearly independent if and only if

$$c_1 \vec{b}_1 + \cdots + c_k \vec{b}_k = 0 \rightarrow c_1 = \cdots = c_k = 0 \tag{2.1}$$

for any real numbers $c_1, \ldots, c_k$; this is only possible if $k \leq n$. Similarly, we
say that a set of $k + 1$ points $\{a_0, a_1, \ldots, a_k\}$ is linearly independent if the
k vectors $\vec{a}_j = a_j - a_0$, $j = 1, \ldots, k$ are linearly independent. Sometimes
we will need to think of points as vectors; in such cases we will identify a
point a with the vector $\vec{a} = a - 0$.

To each vector $\vec{x} \in E^n$ we can associate a rigid motion of translation, a
shift of the entire space in the direction of the vector through a distance
equal to its length. Any set of vectors $\vec{b}_1, \ldots, \vec{b}_k \in E^n$ generates a
countably infinite group under addition (and subtraction) called a Z-

module; its elements are the vectors of the form

$$\vec{x} = m_1\vec{b}_1 + \cdots + m_k\vec{b}_k, \tag{2.2}$$

where the coefficients m_i are integers. We will usually denote an orbit of a Z-module by Ω.

Figure 2.1 shows portions of orbits of three Z-modules, where $n = 2$ and $k = 3$. Although these orbits look very different, they have some crucial features in common.

Definition 2.1 For $x \in \Omega$, the (infinite) star of x is the limit, as $r \to \infty$, of its r-stars (see Definition 1.6).

Proposition 2.1 The stars of Ω are translates of one another.

Proof This follows immediately from the fact that Ω is an orbit of a translation group. ❏

Proposition 2.2 Every point $y \in \Omega$ is a center of symmetry for Ω, and so is the midpoint between y and any other point of Ω.

Proof By Proposition 2.1, we may that assume that $y = 0$. The first statement follows from the fact that Ω contains all integral linear combinations of its points, so $x \in \Omega$ if and only if $-x \in \Omega$. To prove the second, let $x, y, z \in \Omega$. The midpoint of the line segment joining x and y is $\frac{1}{2}(x + y)$; the image of z under inversion through this midpoint is $x + y - z \in \Omega$. ❏

The differences among these orbits are, of course, more obvious than their similarities. The orbit shown in Figure 2.1(a) is a discrete point set, while the point set in (c), if all of it could be shown, would be dense in the plane. In (b) the orbit is stratified in a family of densely filled parallel lines. (Recall that a set A is dense in another set B if every neighborhood of every point of B contains points of A.)

These examples are representative of the general situation: as we will show in Section 2.5, the orbit of a Z-module in E^n *always consists of translates of densely filled d-dimensional subspaces, for some* $d \in \{0, 1, 2, \ldots, n\}$.

Notice that the generators for orbit (a) are both linearly dependent and *integrally* dependent: $5\vec{b}_3 = 3\vec{b}_1 + 4\vec{b}_2$ and thus (2.1) has a solution in integers.

Quasicrystals and geometry

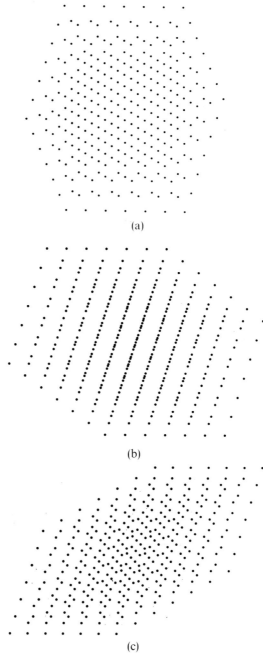

(a)

(b)

(c)

Fig. 2.1 Portions of orbits of groups generated by three vectors in E^2. In all three examples, $\vec{b}_1 = (1, 0)$ and $\vec{b}_2 = (\cos(2\pi/5), \sin(2\pi/5))$. (a) $\vec{b}_3 = \frac{1}{5}(3 + 4\cos(2\pi/5), 4\sin(2\pi/5))$; (b) $\vec{b}_3 = (\cos(4\pi/5), \sin(4\pi/5))$; (c) $\vec{b}_3 = (\sqrt{2}, \sqrt{3})$.

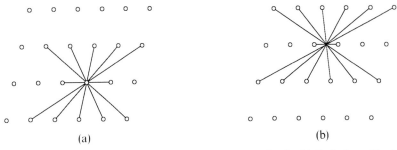

(a)   (b)

Fig. 2.2 (a) Every point of Ω is a center of symmetry for Ω. (b) So is the midpoint between every pair of points.

The number of integrally independent vectors in a module is its *rank*, so in this example the rank is two (in (b) and (c) the rank is three; we postpone discussion of these orbits until Section 2.5).

When the rank of Ω is equal to its span (the dimension of the subspace spanned by $\vec{b}_1, \ldots, \vec{b}_k$), Ω is called a *lattice*.

Definition 2.2 A Z-module in E^n is a lattice (of dimension n) if it is generated by n linearly independent vectors.

Proposition 2.3 A Z-module is a lattice if and only if its orbits are discrete.

Proof This follows immediately from Theorem 2.3 and Proposition 2.13 below.   ❏

We will use the phrase 'point lattice' for an orbit of a Z-module whose generating vectors are linearly independent, reserving the word 'lattice' for the group itself. (While 'lattice' is frequently used for the group or an orbit interchangeably, this can lead to unnecessary confusion.) Lattices will be denoted generically by $\mathcal{L}$, and any orbit of $\mathcal{L}$ by $\mathcal{L}^p$. We can rewrite Proposition 2.3 in the form

Corollary 2.1 An orbit of a Z-module is a point lattice if and only if it is discrete.

2.2 Lattices and their duals

In this section we assume that $k = n$, the dimension of the space in which the vectors lie.

Definition 2.3 Any set of n linearly independent lattice vectors that generates $\mathcal{L}$ is a *basis for $\mathcal{L}$*.

Let $\vec{b}_1, \ldots, \vec{b}_n$ be a basis for $\mathcal{L}$. The squared length or norm $N(\vec{b}_i)$ of $\vec{b}_i$ is

$$N(\vec{b}_i) = |\vec{b}_i|^2 = \vec{b}_i \cdot \vec{b}_i, \qquad (2.3)$$

where $\cdot$ is the ordinary scalar product. Since a point lattice is discrete, $\mathcal{L}$ has a finite set of vectors of minimal (nonzero) norm. These are the *short vectors* of $\mathcal{L}$.

We denote by B the matrix whose columns are the components of the vectors $\vec{b}_i$; B is said to be a *generator matrix* for $\mathcal{L}$. We define $\det \mathcal{L}$, the determinant of $\mathcal{L}$, by $\det \mathcal{L} = \det B$; the determinant is independent of the basis. The symmetric matrix $M = B^T B$, where B^T is the transpose of B, is the *Gramm matrix* for $\mathcal{L}$. Its (i, j)th entry is the scalar product of $\vec{b}_i$ and $\vec{b}_j$; thus its diagonal entries are the norms of the basis vectors. Indeed, M is the matrix of a positive definite quadratic form whose values on $\mathcal{L}$ are the squared lengths of its vectors. If, for all $\vec{x} \in \mathcal{L}$, $N(\vec{x}) \in Z$, then $\mathcal{L}$ is said to be an *integral lattice*.

When M is the $n \times n$ identity matrix I_n, the basis vectors are orthonormal and $\mathcal{L} = \mathcal{I}_n$, the *standard* lattice. ($\mathcal{I}_n$ goes by many other names: the Cartesian lattice, the integer lattice, the primitive hypercubic lattice, and so forth.) Up to scale and orientation, it is the unique n-dimensional lattice with an orthonormal basis; it is an orthogonal direct sum of one-dimensional lattices.

If B is a generator matrix for $\mathcal{L}$, so is BA, where A is any $n \times n$ integer matrix of determinant ± 1 (that is, where A is any element of the 'general linear' group of all unimodular integral matrices, $GL(n, Z)$). Conversely, every matrix that can be written in the form BA is a generator matrix for $\mathcal{L}$. (Be careful: the order of multiplication is important here. *Left* multiplication by A may carry $\mathcal{L}^p$ to a different point lattice.)

Thus every lattice has a countable infinity of different bases. Figure 2.3 shows three bases for the point lattice of Figure 2.1(a).

A *sublattice* $\mathcal{L}' \subset \mathcal{L}$ is any subset that is again a lattice; it may have dimension $k < n$. When $k = n$, the quotient group $\mathcal{L}/\mathcal{L}'$ is finite, and conversely.

Every lattice $\mathcal{L}$ has an associated *dual lattice* $\mathcal{L}^*$, the set of vectors whose scalar products with the vectors of $\mathcal{L}$ are integers.

Definition 2.4 The dual $\mathcal{L}^*$ of $\mathcal{L}$ is the set of vectors $\vec{y} \in E^n$ defined by:

$$\vec{y} \in \mathcal{L}^* \Longleftrightarrow \vec{y} \cdot \vec{x} \in Z \text{ for all } \vec{x} \in \mathcal{L}. \qquad (2.4)$$

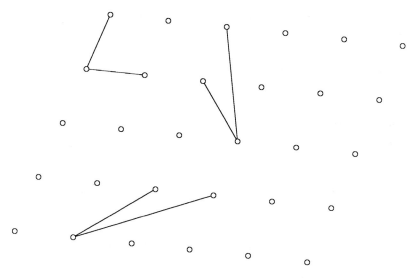

Fig. 2.3 Three bases for the point lattice of Figure 2.1(a).

Dual lattices play a key role in crystallography, particularly in the interpretation of diffraction patterns (Figure 2.4).

Proposition 2.4 $\mathcal{L}^*$ is a lattice.

Proof The n vectors $\vec{b}_1^*, \ldots, \vec{b}_n^*$ defined by the requirement that

$$(\vec{b}_i^*, \vec{b}_j) = \delta_{ij},$$

where δ_{ij} is the Kronecker delta

$$\delta_{jk} = \begin{cases} 1, & \text{if } j = k, \\ 0, & \text{otherwise}, \end{cases}$$

belong to $\mathcal{L}^*$ and are linearly independent. Also, it is easy to show that $\vec{y} \in \mathcal{L}^*$ if and only if it is an integral linear combination of $\vec{b}_1^*, \ldots, \vec{b}_n^*$. □

It follows that if B is a generator matrix for $\mathcal{L}$, then $(B^{-1})^\top$ is a generator matrix for $\mathcal{L}^*$, and the Gramm matrix of $\mathcal{L}^*$ is related to that of $\mathcal{L}$ by the simple formula $M^* = M^{-1}$.

In crystallography, $\mathcal{L}^*$ is usually called the 'reciprocal lattice' because the interplanar spacings of its orbits are inversely proportional to those of $\mathcal{L}$ (as in Figure 2.4). For example – see Appendix II for a sketch of a proof

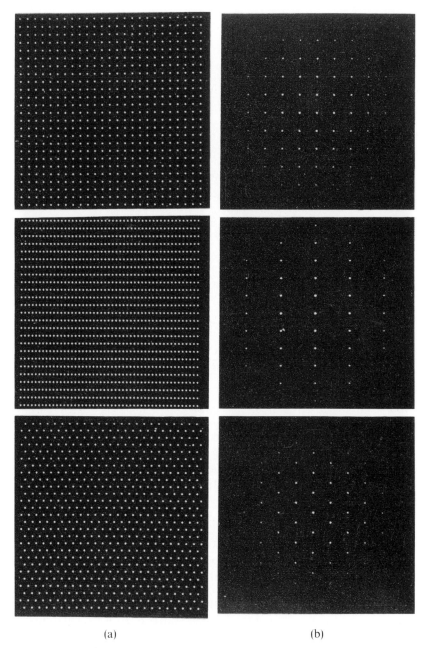

(a)   (b)

Fig. 2.4 (a) Three point lattices. (b) A diffraction pattern is an orbit of the dual lattice. (From *Atlas of Optical Transforms*.)

– if $\mathcal{L}^p$ is a one-dimensional lattice with spacing d between the points of an orbit,

$$\mathcal{L}^p = \{nd \mid n \in Z\}, \tag{2.5}$$

then $\mathcal{L}^{*p}$ is a one-dimensional point lattice with spacing $1/d$:

$$\mathcal{L}^{*p} = \{y \in E^n | y(nd) \in Z\} = \{m/d, \ m \in Z\}. \tag{2.6}$$

It follows immediately from the definition that the dual of $\mathcal{L}^*$ is $\mathcal{L}$:

Proposition 2.5 $(\mathcal{L}^*)^* = \mathcal{L}.$

$\mathcal{L}$ is integral if and only if all its vectors satisfy (2.4), so $\mathcal{L} \subset \mathcal{L}^*$; this is sometimes used as an alternative definition of an integral lattice. As a special case, we have:

Definition 2.5 When $\mathcal{L} = \mathcal{L}^*$, then $\mathcal{L}$ is self dual.

The standard lattice $\mathcal{I}_n$ is both integral and self dual.

Proposition 2.6 Suppose that $\mathcal{L}$ is integral. Then if $\vec{w} \in \mathcal{L}^*$ there exists $j \in Z$ such that $j\vec{w} \in \mathcal{L}$.

Proof $\mathcal{L}$ is a sublattice of finite index in $\mathcal{L}^*$; the proposition follows immediately from the fact that $\mathcal{L}^*/\mathcal{L}$ is a finite abelian group. ❏

From now on, *we will always assume that $\mathcal{L}$ is integral*, unless otherwise stated. (Sometime we will restate this hypothesis, for emphasis.)

2.3 Voronoï cells

In crystallography, lattices are associated with tilings of space by parallelepipedal blocks called *unit cells*, whose edges are parallel to the basis vectors of the lattice. The cell is usually chosen to be *primitive*, which means that its vertices are the only lattice points on its boundary or in its interior; in this case its volume is equal to $|\det \mathcal{L}|$.

Every point lattice $\mathcal{L}^p$ can be partitioned into unit cells in infinitely many ways (one for each choice of basis). Conversely, given one cell, all of $\mathcal{L}^p$ can be reconstructed. The unit cell has limitations as a carrier of visual information, however, because it cannot always be chosen to display the symmetry of the lattice.

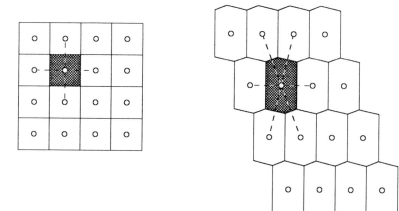

Fig. 2.5 The Voronoï cells of two of the point lattices of Figure 1.13.

Happily, there is another polytope, with the same volume as a unit cell, that does display the symmetry of the lattice and has the further advantage of being independent of the choice of basis. This is the famous *Voronoï* cell. The Voronoï cells for two of the point lattices of Figure 1.13 are shown in Figure 2.5.

Definition 2.6 Let $\Lambda \in E^n$ be any Delone set (or even a finite point set). The Voronoï cell of a point $x \in \Lambda$ is the set of points of E^n that lie at least as close to x as to any other point of Λ:

$$V(x) = \{u \in E^n \mid |x - u| \leq |y - u|, \text{for all } y \in \Lambda\}. \qquad (2.7)$$

To construct the Voronoï cell of x, we connect x to each of the points in its star by straight line segments and construct the ((n-1)-dimensional) perpendicular bisector of each of these segments (Figure 2.6). The Voronoï cell $V(x)$ is then the smallest convex region about x bounded by these hyperplanes.

When Λ is infinite the star of x has infinitely many points and so, in principle, the Voronoï construction calls for an infinite number of operations. But in fact a finite number always suffices (see Chapter 5).

If we carry out this construction for every point of Λ, we obtain a partition of E^n into cells called the *Voronoï tessellation* induced by Λ. It is easy to show that

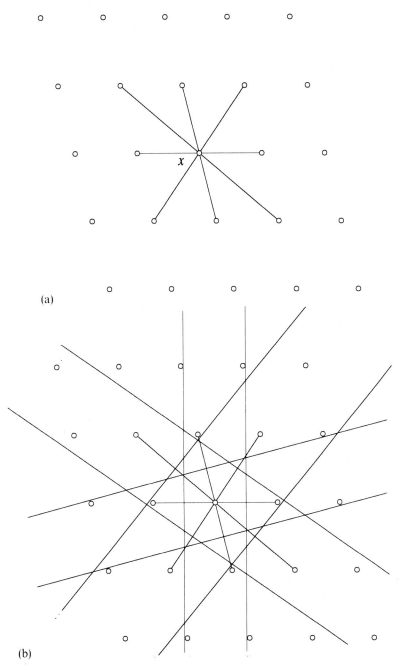

(a)

(b)

Fig. 2.6 The construction of the Voronoï cell of x. (a) Connect x to its neighbors by line segments. (b) Bisect the segments.

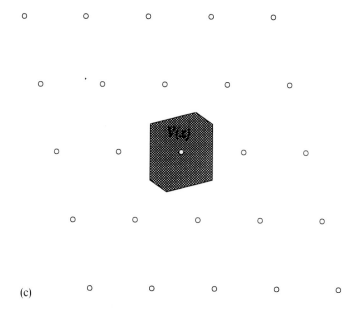

(c)

Fig. 2.6 (c) $V(x)$ is the smallest convex region bounded by the bisectors containing x.

Proposition 2.7 The Voronoï tessellation of a Delone set $\Lambda \subset E^n$ has the following properties:

(i) The cells are convex and fit together along whole faces; no two cells have a common interior point;

(ii) The points of Λ whose Voronoï cells share a vertex v lie on a sphere, centered at v, that has no points of Λ in its interior.

When Λ is the orbit of a group, then the stars of all of its points are congruent and so are their Voronoï cells. In particular, the Voronoï cells of a point lattice are congruent polytopes. (For example, the Voronoï cells of the the points of $\mathcal{I}_n^p$ are n-dimensional unit hypercubes $\mathcal{Q}_n$.) Thus we can speak of the Voronoï cell of a lattice $\mathcal{L}$. Unless otherwise stated, we will assume that 'the' Voronoï cell is $V(0)$. Note that

Proposition 2.8 For all $x \in \mathcal{L}^p$, $V(x) = V(0) + \vec{x}$.

The $(n-1)$-dimensional faces of a Voronoï cell $V(x)$ are its *facets*, and the vectors joining x to another lattice point whose Voronoï cell shares a facet with $V(x)$ are called *facet vectors* (see Figure 2.5). We will denote the set of facet vectors of $\mathcal{L}$ by $F_{\mathcal{L}}$, or simply by F. The set S of short vectors of a point lattice are always facet vectors but except in certain highly symmetric cases (see Section 7) there will be others as well. The Voronoï cell of any point lattice has as least $2n$ facet vectors and at most $2(2^n - 1)$ (Minkowski, 1907).

There is a simple test for determining whether a point of E^n belongs to $V(0)$:

Proposition 2.9 Let $\mathcal{L}$ be a lattice and let $x \in E^n$. Then

$$x \in V(0) \leftrightarrow \vec{x} \cdot \vec{f} \leq \frac{N(\vec{f})}{2}, \text{ for all } \vec{f} \in F_{\mathcal{L}}.$$

The proof follows immediately from Definition 2.6.

The lattice $\mathcal{L}$ is generated by F (but *not* necessarily by S); whether a basis can be extracted from F is an open problem.

Notice that by Proposition 2.2 the Voronoï cell of every lattice point is centrosymmetric. Thus its facets come in parallel pairs, corresponding to pairs of facet vectors $\pm\vec{f}$. It follows that the Voronoï cell of a point lattice is the intersection of a finite number of slabs (a *slab* is a region of E^n bounded by two parallel $(n - 1)$-dimensional hyperplanes). The facets contain the midpoints of the facet vectors corresponding to them and thus, by Proposition 2.2, they are centrosymmetric, too.

The combinatorial structure of a Voronoï cell is sensitively dependent on the lattice parameters. For example, as Figure 2.5 suggests, there are exactly two combinatorial types of Voronoï cell for point lattices in E^2: the rectangle and the hexagon. The Voronoï cell is rectangular if and only if $\mathcal{L}$ has an orthogonal basis; the slightest departure from orthgonality $(\vec{b}_1 \cdot \vec{b}_2 = \epsilon > 0)$ causes the Voronoï cell to become hexagonal.

The five combinatorial types of Voronoï cell for three-dimensional point lattices are Fedorov's parallelohedra (Figure 1.14). There are 52 combinatorial types of Voronoï cells in E^4 (Delone, 1929; Shtogrin, 1973). The exact numbers in higher-dimensional spaces are not known; Engel estimates that there are about 75 000 in E^5 (private communication).

There is no simple relation between the Voronoï cells of $\mathcal{L}$ and $\mathcal{L}^*$; in particular, they are *not* dual polytopes!

2.4 Symmetry

The symmetry group $S(\mathcal{L})$ of a lattice $\mathcal{L}$ is the group of isometries ('rigid motions') that map a point lattice $\mathcal{L}^p$ onto itself; $S(\mathcal{L})$ includes both the translations that carry one lattice point to another and the motions that fix a lattice point. First we will discuss some general properties of isometries, and then we will specialize to those that are possible symmetries of lattices.

2.4.1 Isometries

Definition 2.7 An isometry ϕ of E^n is any transformation that preserves the distances between points: for all $x \in E^n$,

$$|\phi(x - y)| = |x - y|. \tag{2.8}$$

The following propositions are sometimes helpful.

Let ϕ be a mapping of E^n onto itself that keeps the origin fixed. If ϕ preserves distances it preserves angles and vice versa. Thus we have:

Proposition 2.10 Let $\phi : E^n \to E^n$ and $\phi(0) = 0$. Then ϕ is an isometry if and only if $\vec{x} \cdot \vec{y} = \phi(\vec{x}) \cdot \phi(\vec{y})$ for all $\vec{x}, \vec{y} \in E^n$.

Moreover, since an isometry must map a parallelogram with vertices $0, a, b$, and $a + b$ onto a congruent one, we have $\phi(a + b) = \phi(a) + \phi(b)$ and

Proposition 2.11 If ϕ is an isometry and $\phi(0) = 0$, then ϕ is linear.

Definition 2.8 If $\phi(x) = x$, then x is said to be a fixed point of ϕ.

It is useful to distinguish between isometries that keep at least one point fixed and those that do not. We will say that an isometry ϕ that fixes the origin is a *linear isometry*. A linear isometry in E^n can be represented by an $n \times n$ unimodular matrix (det= ± 1), since all of its eigenvalues must have absolute value one. Since det is independent of the choice of basis of E^n, every linear isometry is unambiguously either 'positive' or 'negative' (or in older terminology, 'proper' and 'improper' respectively). An isometry with no fixed points is either a translation or a composite of a translation and an isometry conjugate to a linear isometry. It is easy to see that the set of fixed points of ϕ is a subspace of E^n.

Definition 2.9 The subspace of all points fixed by a linear isometry ϕ is its fixed (sub)space. A proper subspace of E^n that is mapped to itself by ϕ but is not necessarily fixed by it is called a stable subspace.

Linear isometries in spaces of low dimension are easily described in terms of their fixed and stable subspaces. For example, a rotation in E^2 has a single fixed point (the center of rotation) but no proper stable subspace, while a rotation in E^3 has a one-dimensional fixed space (the rotation axis) and a two-dimensional stable subspace (the plane through the origin orthogonal to the axis). A reflection in E^2 has a fixed line and a stable line orthogonal to it; a reflection in E^3 has a fixed plane and a stable line orthogonal to it.

Let ϕ be any one to one linear transformation of E^n onto itself. By iterating ϕ and its inverse, we generate a cyclic group of transformations whose elements are $I, \phi^{\pm 1}, \phi^{\pm 2}, \phi^{\pm 3}, \ldots$. If there is an integer $k \neq 0$ such that $\phi^k = I$ and no smaller integer has this property, then the group has k elements and we say that ϕ is of *order* k. For example, a reflection is an isometry of order two, a rotation through $(2\pi/k)$ radians is an isometry of order k, but a translation is of infinite order.

The geometry of isometries in dimensions $n > 3$ is not as intuitively accessible as when $n = 2$ and $n = 3$, but the following well-known result (see e.g. Engel, 1986) helps us to understand how they work:

Proposition 2.12 By a suitable choice of basis for E^n, any linear isometry can be represented by an $n \times n$ matrix of the form

$$
\mathcal{A} = \begin{pmatrix} A_1 & & & & \\ & A_2 & & & \\ & & A_3 & & \\ & & & \ddots & \\ & & & & A_m \end{pmatrix} \tag{2.9}
$$

where each A_j is either 1,-1, or a 2×2 matrix of the form

$$
\begin{pmatrix} \cos\theta & -\sin\theta \\ \sin\theta & \cos\theta \end{pmatrix} \tag{2.10}
$$

and all other entries are zero.

This matrix describes the action of ϕ on an orthonormal basis of E^n. If $A_j = \pm 1$, then the corresponding basis vector is fixed (+1) or stabilized (−1), while if A_j is a nondiagonal 2×2 matrix, then two basis vectors define a plane that is stabilized (but not fixed) by the operation ϕ.

Rather than think up new names for the many different kinds of isometry in E^n, we will simply say that ϕ is a rotation if and only if $\det A > 0$. For example, if we include the cases $\theta = 0$, $\phi = 0$, and

$\theta = \phi = 0$, every rotation in E^4 (that fixes the origin) can be written in the form

$$\begin{pmatrix} \cos\theta & -\sin\theta & 0 & 0 \\ \sin\theta & \cos\theta & 0 & 0 \\ 0 & 0 & \cos\phi & -\sin\phi \\ 0 & 0 & \sin\phi & \cos\phi \end{pmatrix}, \qquad (2.11)$$

where ϕ and θ are the angles of rotation. The two planes are stable sets; they are the eigenspaces of the rotation. The fixed point set is just the origin.

The block structure of $\mathcal{A}$ is preserved when the matrix is raised to a power:

$$\mathcal{A}^k = \begin{pmatrix} A_1^k & & & & \\ & A_2^k & & & \\ & & A_3^k & & \\ & & & \ddots & \\ & & & & A_m^k \end{pmatrix}. \qquad (2.12)$$

Thus the order of ϕ is the least common multiple of the orders of $A_1, \ldots, A_m$.

2.4.2 Symmetry groups

Definition 2.10 The symmetry group of an object is the maximal group of isometries that stabilizes it.

Although every linear isometry can be written in the form of (2.9) with respect to *some* basis, it need *not* be true that all the isometries in the symmetry group of an object can be written in this form with respect to the *same* basis. When all the stabilizers of an object can be written in the same block form with respect to a single basis, say

$$\begin{pmatrix} B_1 & 0 \\ 0 & B_2 \end{pmatrix}, \qquad (2.13)$$

where B_1 is $k \times k$, B_2 is $m \times m$, and $k > 0, m = n - k > 0$, then the symmetry group is said to be *reducible*; otherwise it is *irreducible*. The fixed set of an irreducible group is just the origin; irreducible groups have no proper stable sets.

The linear isometries of E^n form the *orthogonal group* $O(n)$. It is helpful to think of $O(n)$ as the symmetry group of the n-dimensional unit sphere centered at the origin. The finite subgroups of $O(2)$ were classified long ago (no one knows by whom, although Leonardo da Vinci is sometimes credited with their discovery). In $O(2)$ there are two infinite families of groups, the cyclic rotation groups C_n and the dihedral groups D_n, the symmetry groups of rosettes. The reducible subgroups of $O(3)$ have stable planes and stable or fixed lines orthogonal to them. There are several infinite families; all of them can be regarded as generalizations of C_n and D_n (they are parameterized by the order n of a rotation that is unique when $n > 2$). The family of irreducible groups is finite: it contains only the symmetry groups of the regular polyhedra and certain of their subgroups. (For more details on the finite isometry groups, see Senechal, 1991(c).)

The symmetry group G of a regular system of points (see Definition 1.7) is a crystallographic group. It follows from Theorem 1.2 that G has a maximal abelian subgroup T of translations of finite index and the 'system' is a union of a finite number of point lattices. The quotient group G/T is isomorphic to a finite group P of linear isometries that map T onto itself; each element of P corresponds to a different isometry. Since any group of linear isometries has a fixed point (the origin), P is usually called the *point group* of the lattice. Crystallographic groups are thus 'products' of T and P, but are *never* direct products: P is not a normal subgroup, since it is conjugate to all of the groups that fix the lattice points. In special cases, including the case where the regular system of points is a point lattice, G is a *semidirect* product of T and P.

The points of a lattice have integer coordinates with respect to any basis $\vec{b}_1, \ldots, \vec{b}_n$ of $\mathcal{L}$; it follows that P can be represented by a finite group of unimodular matrices with integer entries. In other words, P can be identified with a finite subgroup of $GL(n, Z)$.

Although every element of P maps the lattice onto itself, neither its fixed point set nor its stable sets need contain any lattice points other than the origin.

Definition 2.11 Let $\mathcal{L}$ be a lattice and $\mathcal{E}$ any k-dimensional subspace of E^n, where $0 < k < n$. If $\mathcal{E} \cap \mathcal{L} = \{0\}$ then $\mathcal{E}$ is said to be totally irrational.

By Proposition 2.2, P always has at least two elements, the identity isometry I_n and also the isometry $-I_n$, which maps every point to its image through the origin, $x \to -x$. Most lattices (indeed, all except a set of measure zero) have no other point symmetries. However, all of the lattices that have been employed in quasicrystal theory (so far) are highly symmetrical.

Proposition 2.13 $\mathcal{L}$ and $\mathcal{L}^*$ are invariant under the same isometries.

Proof Let ϕ be an isometry that maps $\mathcal{L}$ to itself. We need to show that if $\vec{y} \in \mathcal{L}^*$ then $\phi\vec{y} \in \mathcal{L}^*$. For $\vec{x} \in \mathcal{L}$ and $\vec{y} \in \mathcal{L}^*$ we have, by Proposition 2.10,

$$\phi\vec{y} \cdot \vec{x} = \phi\vec{y} \cdot \phi\phi^{-1}\vec{x} = \vec{y} \cdot \phi^{-1}\vec{x} \in Z \tag{2.14}$$

since $\phi^{-1}\vec{x} \in \mathcal{L}$. Thus $\mathcal{L}^*$ is also invariant under ϕ. ❏

It follows that the point groups P and P^* of $\mathcal{L}$ and $\mathcal{L}^*$ are isomorphic, but this does not mean that P and P^* can be represented by conjugate subgroups of $GL(n, Z)$: they may have inequivalent integer representations.

2.4.3 The crystallographic restriction

Using matrices, we can give another proof of the crumbling cornerstone of mathematical crystallography. Unlike the geometrical argument in Chapter 1, this algebraic one can be generalized to higher dimensions.

Theorem 2.1 (the crystallographic restriction). If R is an element of the point group of a two- or three-dimensional lattice $\mathcal{L}$, then its order is two, three, four, or six.

Proof R can be represented both by a matrix of the form (2.9) and by an element of $GL(n, Z)$; since these matrices represent the same motion with respect to different bases, they are conjugate. Conjugate matrices have the same trace, and the traces of elements of $GL(n, Z)$ are integers since the entries of the matrices are. This means that when $n = 2$, we must have

$$2\cos\theta \in Z . \tag{2.15}$$

But $|\cos(\theta)| \leq 1$ and so $\cos(\theta) \in \{0, \pm\frac{1}{2}, \pm 1\}$. Assuming that $0 \leq \theta < 2\pi$, this is possible only for

$$\theta \in \{0, \pi, \pi/2, 2\pi/3, 2\pi/6\}, \tag{2.16}$$

which proves the theorem. (It is easy to modify this argument to include the case where $n = 3$.) ❏

Applying the restriction to the list of finite subgroups of $O(3)$, we see that all but a small number of the groups in the infinite families are eliminated, and so are the icosahedral groups. We are left with only 32 groups, of which seven can serve as point groups for lattices (the seven 'crystal systems').

The other point groups contain rotations of order 5 or 7, 8,... They used to be called *noncrystallographic*; now we need a better name for them. How about *quasicrystallographic?*

There are only two finite quasicrystallographic irreducible subgroups of $O(3)$: the full icosahedral group and its subgroup of rotations. *Thus three-dimensional quasicrystals must be either 'icosahedral' or 'essentially two-dimensional'.*

To state the analogue of the crystallographic restriction for lattices in E^n, let $n(k)$ be the least value of n for which an element of order k appears in $GL(n, Z)$. Recall *Euler's ϕ function*, defined for the positive integers: $\phi(n)$ is the number of integers less than and relatively prime to n (to refresh your memory, see Appendix I). Now define

$$\Phi(n) = \begin{cases} \phi(n) & \text{if } n = p^\alpha, \text{ where } p \text{ is a prime and } \alpha \in N; \\ \Phi(n_1) + \Phi(n_2) & \text{if } n = n_1 n_2 \text{ and } (n_1, n_2) = 1. \end{cases}$$

Here $(n_1, n_2) = 1$ means that n_1 and n_2 are relatively prime.

Theorem 2.2 $n(k) = \Phi(k)$.

Thus rotations of order 5 first appear in E^4. For two different proofs of this theorem, see Hiller (1985) and Senechal (1992). Hiller's paper also contains a table of values of $n(k)$.

When $n = 4$ we also encounter lattices whose point groups contain elements of order 8, 10, or 12. For example, the standard lattice $\mathcal{I}_4 \in E^4$ is invariant under a rotation of order 8. This lattice has eight short vectors of norm 1: $(\pm 1, 0, 0, 0)$, $(0, \pm 1, 0, 0)$, $(0, 0, \pm 1, 0)$, and $(0, 0, 0, \pm 1)$; they are precisely the facet vectors of the Voronoï cell. The point group P of $\mathcal{I}_4$ is the symmetry group of the hypercube $\mathcal{Q}_4$; it includes the element

$$\begin{pmatrix} 0 & 0 & 0 & -1 \\ 1 & 0 & 0 & 0 \\ 0 & 1 & 0 & 0 \\ 0 & 0 & 1 & 0 \end{pmatrix}, \tag{2.17}$$

which is a rotation (since its determinant is positive) of order 8. More generally, the point group of $\mathcal{I}_n$ always contains an element of order $2n$.

Icosahedral symmetry – with its rotations of orders 5, 3, and 2 – becomes possible for a lattice only when $n \geq 6$.

2.5 Orbits of Z-modules

Our claim (in Section 2.1) that the orbits in Figure 2.1 show all the possibilities for $n = 2$ and $k = 3$ is based on the following fundamental theorem.

Theorem 2.3 Let $\phi : R^k \to R^n$ be a surjective linear mapping, where $n < k$. Then there are (possibly trivial) subspaces V, W of R^n such that $R^n = V \oplus W$ and

  (i) $\phi(Z^k) = \phi(Z^k) \cap W + \phi(Z^k) \cap V$,
  (ii) $\phi(Z^k) \cap W$ is a discrete point lattice in W,
  (iii) $\phi(Z^k) \cap V$ is a dense subgroup of V.

Thus for the orbit shown in Figure 2.1(a), $V = \{0\}$ and dim $W = 2$. Here $\vec{b}_3$ is a rational linear combination of $\vec{b}_1$ and $\vec{b}_2$, and an integral linear combination of $\frac{1}{5}\vec{b}_1$ and $\frac{1}{5}\vec{b}_2$. In fact the Z-module generated by $\vec{b}_1, \vec{b}_2, \vec{b}_3$ is precisely the lattice generated by $\frac{1}{5}\vec{b}_1$ and $\frac{1}{5}\vec{b}_2$. In (b), dim $V =$ dim $W = 1$ because b_3 is the sum of a rational multiple of b_1 and an irrational multiple of b_2. In (c), dim $V = 2$ while $W = \{0\}$; this is due to the fact that the coefficients of $\vec{b}_3$ with respect to the $\vec{b}_1, \vec{b}_2$ basis are rationally independent irrational numbers. The theorem is proved in Appendix I.

Proposition 2.14 Every orbit Ω of a Z-module in E^n is the orthogonal projection of a point lattice in E^k onto E^n, where $k \geq n$.

Proof The following straightforward algorithm shows one way (among many) to construct a point lattice that projects to Ω. Let

$$\vec{b}_1, \ldots, \vec{b}_n, \vec{b}_{n+1}, \ldots, \vec{b}_k$$

be the generators of Ω, and assume that $\vec{b}_1, \ldots, \vec{b}_n$ span E^n, the space in which the k vectors lie. Embed $\mathcal{E} = E^n$ as an n-dimensional subspace in E^k. We can choose an orthonormal basis $\vec{a}_1, \ldots, \vec{a}_k$ for E^k such that $\vec{a}_{n+1}, \ldots, \vec{a}_k \in \mathcal{E}^{\perp}$, the orthogonal complement of $\mathcal{E}$. Now just lift the vectors $\vec{b}_j$, $j = n + 1, \ldots, k$ to linearly independent vectors $\vec{b}'_{n+1}, \ldots, \vec{b}'_k$ in E^k, for example by adding some multiple of $\vec{a}_j$ to $\vec{b}_j$. ❑

Although Ω can always be lifted to *some* point lattice, it is not always the case that it can be lifted to a point lattice with prescribed properties. In particular, it is not always possible to lift it to a standard point lattice. A theorem of Hadwiger tells us precisely when this can be done. An orthogonal projection from E^k to E^n of $2k$ vectors $\pm\vec{a}_1,\ldots,\pm\vec{a}_k$, where the $\vec{a}_1,\ldots,\vec{a}_k$ are pairwise orthogonal and have the same length, is called a *eutactic star.*

Theorem 2.4 (Hadwiger) Let $\vec{b}_1,\ldots,\vec{b}_k \in E^n$, $n \leq k$. Let B be the $n \times k$ matrix whose jth column is the vector $\vec{b}_j$. Then $\pm\vec{b}_1,\ldots,\pm\vec{b}_k$ is a eutactic star if and only if

$$B B^T = \lambda I_n,$$

where I_n is the $n \times n$ identity matrix and λ is a real scalar.

If $\lambda = 1$, then $\pm\vec{b}_1,\ldots,\pm\vec{b}_k$ lift to an orthonormal frame or 'cross'. For a proof of this theorem, see Hadwiger (1940) or Coxeter (1973).

For example, if $n = 1$ and $\vec{b}_1,\vec{b}_2,\ldots,\vec{b}_k \in E^1$ then $\vec{b}_j = c_j\vec{b}_1$ for some real numbers $c_j, j = 2,\ldots,k$ and $B = (1, c_2,\ldots,c_k)$ is the $1 \times k$ matrix of generators. We have

$$B B^\top = 1 + \sum_{j=2}^{k} c_j^2,$$

which is a multiple of the identity in E^1. Thus *any* finite set of vectors in E^1 can be lifted to an orthogonal frame in E^k.

As a second example, let $n = 2$ and let the k vectors $\vec{b}_1,\ldots,\vec{b}_k$, $k \geq 3$, point to the vertices of a regular k-gon:

$$\vec{b}_j = (\cos(2j\pi/k), \sin(2j\pi/k)), \quad j = 0,\ldots,k-1. \tag{2.18}$$

These vectors are certainly not linearly independent; indeed, their sum is 0. However, they constitute a eutactic star. We have

$$B = \begin{pmatrix} 1 & \cos(2\pi/k) & \ldots & \cos(2(k-1)\pi/k) \\ 0 & \sin(2\pi/k) & \ldots & \sin(2(k-1)\pi/k) \end{pmatrix};$$

to show that $B B^T = \lambda I_2$ we need to show that

$$\sum_{j=1}^{k} \cos^2(2j\pi/k) = \sum_{j=1}^{k} \sin^2(2j\pi/k), \quad \sum_{j=1}^{k} \cos(2j\pi/k)\sin(2j\pi/k) = 0.$$
$$\tag{2.19}$$

This is a straightforward exercise in juggling trigonometric identities.

In a similar way, we can show that the set of 12 vectors joining the center of a regular icosahedron to its vertices is the projection into E^3 of an orthonormal cross in E^6.

2.6 High-dimensional lattices, low-dimensional crystallography

The Delone set of Figure 2.7 is not a point lattice, nor is it the orbit of any
Z-module whose rank is greater than its span. But, much as the ancients
'connected the dots' to draw fantastic designs in the sky, we can
superimpose a pattern on Figure 2.7 with 'celestial' order (Figure 2.8)!

Figure 2.8 shows that every point in Figure 2.7 belongs to an orbit of
the Z module generated by five vectors pointing to the vertices of a regular
pentagon. But the set we see is a discrete sub*set* of the orbit; it is not itself
the orbit of any group. You may also recognize Figure 2.8 as a portion of
a Penrose tiling of the plane.

2.6.1 *The projection method*

The point set shown in Figure 2.7 – and the Penrose tiling – can be
obtained by projecting a subset of a point lattice onto a plane. Projection

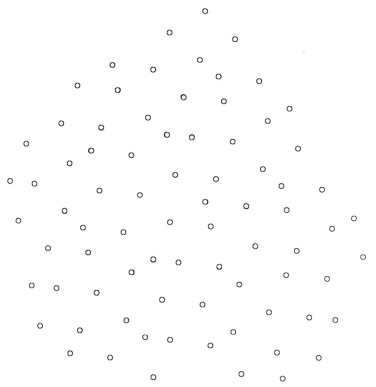

Fig. 2.7 Is any pattern lurking here?

is a powerful technique for constructing nonperiodic patterns and tilings, independently of any connection with crystals. We will describe the special case called the 'canonical' projection. Let $\mathcal{E}$ be a totally irrational d-dimensional subspace of E^n, and let $\mathcal{E}^\perp$ be its orthogonal complement ($\mathcal{E}^\perp$ may or may not be totally irrational). Let Π be the orthogonal projector onto $\mathcal{E}$, and $\Pi^\perp$ the orthogonal projector onto $\mathcal{E}^\perp$. Since Π and $\Pi^\perp$ are linear maps, $\Pi(\mathcal{L}^p)$ and $\Pi^\perp(\mathcal{L}^p)$ are orbits of Z-modules in $\mathcal{E}$ and $\mathcal{E}^\perp$, respectively, generated by the projections of any (every) basis of $\mathcal{L}^p$.

Proposition 2.15 When $\mathcal{L}$ is an integral lattice, then the following statements are equivalent:

  (i) $\Pi(\mathcal{L}^p)$ is everywhere dense in $\mathcal{E}$;
  (ii) $\mathcal{E} \cap \mathcal{L}^p = \{0\}$;
  (iii) $\Pi^\perp|_{\mathcal{L}^p}$ is one to one.

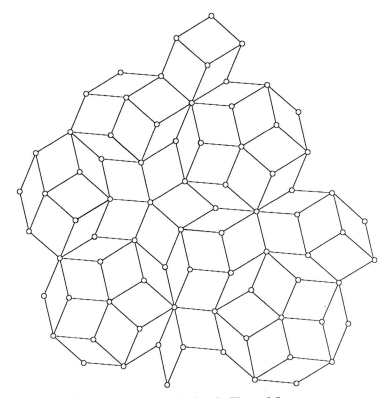

Fig. 2.8 Connecting the dots in Figure 2.7.

Proof First we show that (i) ↔ (ii). We have (i) → (ii) because, if $\vec{w} \in \mathcal{E} \cap \mathcal{L}$, $\vec{w} \neq 0$, then $\vec{w} \in \mathcal{E} \cap \mathcal{L}^*$ and the subspace W of Theorem 2.3 is nontrivial. Conversely, if W is nontrivial then there is a nonzero vector $\vec{w} \in \mathcal{E}$ such that

$$\forall x \in \mathcal{L}, \quad \vec{w} \cdot \Pi(\vec{x}) \in Z. \tag{2.20}$$

Since $\vec{x} = \Pi(\vec{x}) + \Pi^\perp(\vec{x})$ and $\vec{w} \in \mathcal{E}$, we have

$$\vec{w} \cdot \vec{x} = \vec{w} \cdot \Pi(\vec{x}) + 0. \tag{2.21}$$

Thus $\vec{w}$ satisfies (2.20) if and only if $\vec{w} \in \mathcal{L}^*$ and, by Proposition 2.6, $\mathcal{E} \cap \mathcal{L} \neq \{0\}$. The equivalence of (ii) and (iii) is immediate, since $\mathcal{E}$ is the kernel of $\Pi^\perp$. ❏

(When $\mathcal{L}$ is not integral, the density of $\Pi(\mathcal{L}^p)$ is equivalent to the condition $\mathcal{L}^* \cap \mathcal{E} = \{0\}$.)

Since $\Pi(\mathcal{L}^p)$ is dense in $\mathcal{E}$, it is not a Delone set. To obtain one, we must select a subset of the points of $\mathcal{L}^p$ for projection.

The first step is to fix a compact subset $K \subset \mathcal{E}^\perp$ with nonempty interior; K will be called the *window*, or acceptance domain, for the projection. We will project, onto $\mathcal{E}$, those points $x \in \mathcal{L}^p$ such that $\Pi^\perp(x) \in K$. These are the points that lie in the cylinder $C = K \oplus \mathcal{E}$. Let $X = C \cap \mathcal{L}^p$ (Figure 2.9).

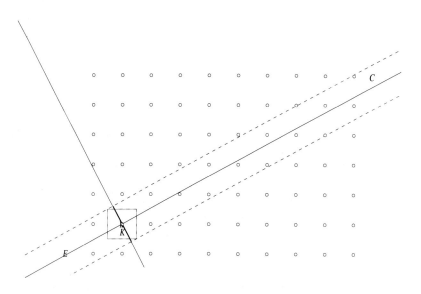

Fig. 2.9 K is the window and C is the cylinder $K \oplus \mathcal{E}$; $X = C \cap \mathcal{L}^p$.

Proposition 2.16 $\Pi(X)$ is a Delone set.

Proof We must show that $\Pi(X)$ is discrete and relatively dense in $\mathcal{E}$. To prove discreteness, it is sufficient to show that there is a neighborhood of the origin in $\mathcal{E}$ that contains no other points of $\Pi(X)$. Let $x \in \mathcal{L}^p$ and assume that $\Pi(x)$ lies in $B_0(c)$, the ball of radius $c > 0$ about the origin. Since

$$|x|^2 = |\Pi(x)|^2 + |\Pi^\perp(x)|^2$$

and the set $\Pi^\perp(X)$ is bounded by assumption, x must lie in a sphere about 0 of some finite radius $m > 0$. Since $\mathcal{L}^p$ is discrete, $\mathcal{L}^p \cap B_0(m)$ is a finite subset $U \subset L^p$. Then $\Pi(U)$ is also finite and for sufficiently small $r > 0$ we have $\Pi(U) \cap B_0(r) = \{0\}$. Relative density is immediate because $\mathcal{L}^p$ is relatively dense in E^n and hence in the cylinder C. ❑

Next we show that this Delone set is nonperiodic.

Proposition 2.17 If $\mathcal{E}$ is totally irrational, then $\Pi(X)$ is nonperiodic.

Proof We decompose $\mathcal{E}^\perp$ into subspaces V and W, as in Theorem 2.3. Since $\Pi^\perp(\mathcal{L})$ is one-to-one and dim $\mathcal{E}^\perp < n$, V is nontrivial, and the restriction of Π to $(V \oplus \mathcal{E}) \cap \mathcal{L}$ is one-to-one since $\Pi^\perp((V \oplus \mathcal{E}) \cap \mathcal{L})$ is dense in V. Thus, since K has nonempty interior, $(K \cap V) \cap \Pi^\perp(\mathcal{L})$ is dense in $K \cap V$. Write $X_V = ((K \cap V) \oplus \mathcal{E}) \cap \mathcal{L}$. It is easy to show that if $\Pi(X_V)$ is periodic then X_V is. However, X_V cannot be periodic because K is compact and thus cannot contain all of the integer multiples of any vector. Since $(K \cap W) \cap \Pi^\perp(\mathcal{L})$ is *finite*, X, and thus $\Pi(X)$, cannot be periodic either. ❑

Thus we have a simple but powerful way of constructing nonperiodic Delone sets by projecting subsets of higher-dimensional lattices:

Selection criterion. Project $x \in \mathcal{L}$ orthogonally onto $\mathcal{E}$ if and only if $\Pi^\perp(x)$ lies in the window K.

The 'canonical' choice for the window is $\Pi^\perp(V(0))$, where $V(0)$ is the Voronoï cell of the origin of $\mathcal{L}$. Figure 2.10 shows how canonical projection works when $k = 2$ and $d = 1$. Canonical projection can also be used to construct tilings of $\mathcal{E}$ (see Chapters 4 and 5).

The selection criterion, for canonical projection, can be expressed in several equivalent ways.

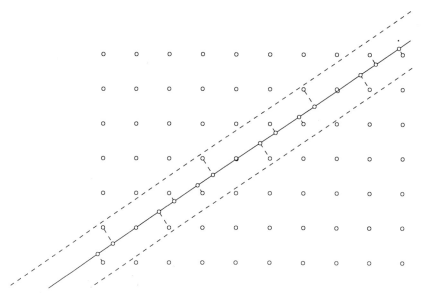

Fig. 2.10 Canonical projection.

Proposition 2.18 For canonical projection $(K = \Pi^\perp(V(0)))$ the following statements are equivalent.

(i) $x \in X$;
(ii) $\Pi^\perp(x) \in \Pi^\perp(V(0))$;
(iii) $\mathcal{E} \cap V(x) \neq \emptyset$.

Proof The equivalence of (i) and (ii) follows immediately from the construction of X above. To show that (ii) and (iii) are equivalent, we note that $\Pi^\perp(x) \in \Pi^\perp(V(0)) \leftrightarrow \exists \vec{e} \in \mathcal{E}$ such that $x + \vec{e} \in V(0) \leftrightarrow e \in V(x)$.

◻

2.6.2 Uniform distribution

The Delone set we have just constructed does not have the properties of Propositions 2.1 and 2.2; indeed, its points cannot have congruent stars, because then, by Bieberbach's theorem (Theorem 1.2) it would have translational symmetry. However, we will see in Chapters 4 and 5 that the projection method guarantees that there is always a finite number of 'local' stars of any radius; both the shapes of the stars and their relative frequencies are determined by the window. This important result depends on the fact that the points of $\Pi^{\perp}(\mathcal{L})$ are uniformly distributed in the densely filled subspaces of $\mathcal{E}^{\perp}$. Uniform distribution is a fundamental property of projected lattices that is closely related to the classical definition of uniform distribution of sequences (see Appendix I).

To explain this notion, let $\mathcal{L}$ be any lattice in E^n, not necessarily integral, and let us assume that $\Pi^{\perp}(\mathcal{L})$ is dense in $\mathcal{E}^{\perp}$; the cases where it is not can be handled in a straightforward way. Let K_1 and K_2 be any two $(n-d)$-dimensional cubes in $\mathcal{E}^{\perp}$ and let $J \subset \mathcal{E}$ be a d-dimensional cube centered at the origin. For any positive real number λ, we set

$$P_{\lambda}^1 = K_1 \oplus \lambda J, \quad P_{\lambda}^2 = K_2 \oplus \lambda J.$$

Then P_{λ}^1 and P_{λ}^2 are bounded cylinders in E^n with 'bases' K_1 and K_2. Note that $\Pi^{\perp}(P_{\lambda}^1 \cap \mathcal{L})$ is a finite subset of K_1, and similarly for K_2.

'Uniform distribution' of $\Pi^{\perp}(\mathcal{L})$ in $\mathcal{E}^{\perp}$ means that, for sufficiently large λ, the ratio of the number of points of $\Pi^{\perp}(P_{\lambda}^1 \cap \mathcal{L})$ and $\Pi^{\perp}(P_{\lambda}^2 \cap \mathcal{L})$ in K_1 and K_2 is approximately equal to the ratio of the volumes of K_1 and K_2:

$$\lim_{\lambda \to \infty} \frac{|P_{\lambda}^1 \cap \mathcal{L}|}{|P_{\lambda}^2 \cap \mathcal{L}|} = \frac{Vol(K_1)}{Vol(K_2)}.$$

It is somewhat tedious but not difficult to show (de Bruijn and Senechal, 1995) that this is always the case when $\Pi^{\perp}(\mathcal{L})$ is dense in $\mathcal{E}^{\perp}$. This can be generalized in various ways, for example the Ks can be any polytopes.

Among other things, uniform distribution implies that if we translate the cylinder used in a projection, the pattern we obtain by projecting the lattice points in the translate will belong to the same local isomorphism class as the original pattern (Figure 2.11). We will also use uniform distribution in Chapter 4 to compute the diffraction spectra of projected crystals.

2.6.3 *Five-fold symmetry*

In quasicrystal theory, $\mathcal{E}$ is chosen to be a subspace that is stable under some subgroup of the point group of $\mathcal{L}$. Then $\Pi(X)$ exhibits some 'traces' of the symmetry of this point group: this symmetry will appear locally but not globally. However, the global symmetry will reappear in diffraction patterns of the projected set.

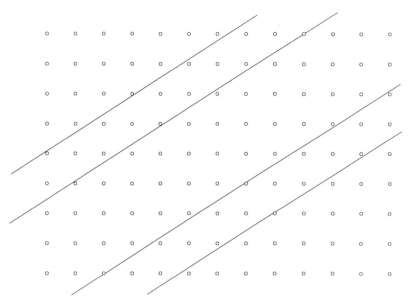

Fig. 2.11 Project and compare!

As a specific example of the projection method, we will construct a Delone set with local five-fold symmetry by projecting a subset of $\mathcal{I}_5$ onto a plane. This point set will be closely related to the vertex set of a Penrose tiling. (The subtle modifications needed to obtain the Penrose tilings will emerge in later chapters.)

The first step is to construct the subspace $\mathcal{E}$. The Voronoï cell of $\mathcal{I}_5$ is a five-dimensional unit hypercube $\mathcal{Q}_5$, which has ten four-dimensional facets, with facet vectors

$$\pm\vec{e}_1 = (\pm 1, 0, 0, 0, 0), \ldots, \pm\vec{e}_5 = (0, 0, 0, 0, \pm 1)$$

and 32 vertices:

$$\sum_{j=1}^{5} \alpha_j \vec{e}_j, \quad \alpha_j \in \{-\frac{1}{2}, \frac{1}{2}\}.$$

The 'body diagonal' of $\mathcal{Q}_5$ is the line segment from $(0, 0, 0, 0, 0)$ to $(1, 1, 1, 1, 1)$. Figure 2.12 shows the projection of the hypercube onto a plane orthogonal to $\vec{w} = (1, 1, 1, 1, 1)$. (The two vertices mapped to the center of the diagram are identified under this projection.)

Five-fold rotation ϕ about $\vec{w}$ is a symmetry of the hypercube, since ϕ permutes the five basis vectors $\vec{e}_j$, $j = 1, \ldots, 5$ cyclically. With respect to

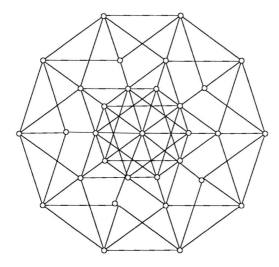

Fig. 2.12 The hypercube $\mathcal{Q}_5$ projected orthogonally to a plane.

this basis, the rotation matrix is

$$\mathcal{A} = \begin{pmatrix} 0 & 0 & 0 & 0 & 1 \\ 1 & 0 & 0 & 0 & 0 \\ 0 & 1 & 0 & 0 & 0 \\ 0 & 0 & 1 & 0 & 0 \\ 0 & 0 & 0 & 1 & 0 \end{pmatrix}.$$

Let us also construct a basis for E^5 with respect to which we can write $\mathcal{A}$ in the form of (2.9). The eigenvalues of $\mathcal{A}$ are the fifth roots of unity,

$$1, \xi, \xi^2, \xi^3, \xi^4,$$

where

$$\xi^j = \exp(2\pi i j/5), \quad j = 0, \ldots, 4 \text{ and } \xi^3 = \xi^{-2}, \xi^4 = \xi^{-1}.$$

The only real eigenvalue is 1; the others are complex and occur in conjugate pairs, ξ, ξ^{-1} and ξ^2, ξ^{-2}. This means that E^5 decomposes into *two* two-dimensional planes, both stable under ϕ, both totally irrational, and both orthogonal to the fixed space Δ generated by $\vec{w}$ (and to one another). The vectors $\vec{e}_j$ project to the vertices of a regular pentagon in both planes, but in one of them (say, $\mathcal{E}$) the angle between the projections of $\vec{e}_j$ and $\vec{e}_{j+1}$ is $2\pi/5$ while in the other, $\mathcal{E}'$, it is $4\pi/5$ (Figure 2.13). We have $\mathcal{E}^\perp = \mathcal{E}' \oplus \Delta$. Notice that $\mathcal{E}^\perp$ is *not* totally irrational, since Δ

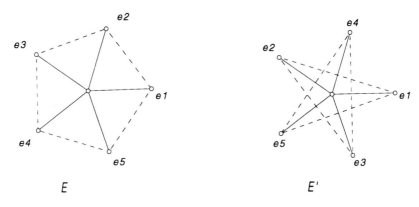

Fig. 2.13 Projections of $\vec{e}_j, j = 1, \ldots, 5$, onto $\mathcal{E}$ and $\mathcal{E}'$.

contains the multiples of $\vec{w}$. Indeed, $\Pi^\perp(\mathcal{L}^p)$ does not fill $\mathcal{E}^\perp$ densely: the projected orbit consists of translates of densely filled planes. This is a special case of Theorem 2.3, where $n = 3, V = 2, W = 1$.

We compute eigenvectors belonging to these eigenvalues in the usual way. Obviously, $\vec{w}$ is an eigenvector belonging to the eigenvalue 1. The eigenvectors belonging to ξ^j, $j = 1, \ldots, 4$ are multiples of

$$\vec{v}_j = (1, \xi^j, \xi^{2j}, \xi^{3j}, \xi^{4j}) \ .$$

Note that

$$\vec{w} \cdot \vec{v}_j = 0, \ j = 1, \ldots, 4$$

and also

$$\vec{v}_1 \cdot \vec{v}_2 = \vec{v}_1 \cdot \vec{v}_3 = \vec{v}_4 \cdot \vec{v}_2 = \vec{v}_4 \cdot \vec{v}_3 = 0 \ .$$

Thus $\vec{v}_1$ and $\vec{v}_4$ span a plane orthogonal to the plane spanned by $\vec{v}_2$ and $\vec{v}_3$, and both planes are orthogonal to $\vec{w}$. Although $\vec{v}_1, \ldots, \vec{v}_4$ have complex entries, they define *real* planes. The $\vec{v}_1, \vec{v}_4$ plane contains

$$\vec{v}_1 + \vec{v}_4 = (1, \exp(2\pi i/5) + \exp(-2\pi i/5), \ldots, \exp(8\pi i/5) + \exp(-8\pi i/5))$$

$$= 2(1, \cos(2\pi/5), \ldots, \cos(8\pi/5))$$

and

$$i(\vec{v}_1 - \vec{v}_4) = i(0, \exp(2\pi i/5) - \exp(-2\pi i/5), \ldots,$$

$$\exp(8\pi i/5) - \exp(-8\pi i/5))$$

$$= 2(0, \sin(2\pi/5), \ldots, \sin(8\pi/5)).$$

Notice that

$$(\vec{v}_1 + \vec{v}_4) \cdot i(\vec{v}_1 - \vec{v}_4) = 4 \sum_{j=0}^{4} \cos(2j\pi/5) \sin(2j\pi/5) = 2 \sum_{j=0}^{4} \sin(4j\pi/5) = 0.$$

Similarly, the $\vec{v}_2, \vec{v}_3$ plane contains two real orthogonal vectors,

$$\vec{v}_2 + \vec{v}_3 = 2(1, \cos(4\pi/5), \ldots, \cos(6\pi/5))$$

and

$$i(\vec{v}_2 - \vec{v}_3) = 2(0, \sin(4\pi/5), \ldots, \sin(6\pi/5)).$$

Thus letting $\vec{u}_1, \vec{u}_2, \vec{u}_3, \vec{u}_4$ be the unit vectors parallel to $\vec{v}_1 + \vec{v}_4$, $i(\vec{v}_1 - \vec{v}_4), \vec{v}_2 + \vec{v}_3$ and $i(\vec{v}_2 - \vec{v}_3)$, and setting $\vec{u}_5 = \frac{1}{\sqrt{5}}\vec{w}$, we have an orthonormal basis of eigenvectors of the rotation ϕ. With respect to the basis $\vec{u}_1, \ldots, \vec{u}_5$, the matrix for ϕ is

$$\begin{pmatrix} \cos(2\pi/5) & -\sin(2\pi/5) & 0 & 0 & 0 \\ \sin(2\pi/5) & \cos(2\pi/5) & 0 & 0 & 0 \\ 0 & 0 & \cos(4\pi/5) & -\sin(4\pi/5) & 0 \\ 0 & 0 & \sin(4\pi/5) & \cos(4\pi/5) & 0 \\ 0 & 0 & 0 & 0 & 1 \end{pmatrix}.$$

The matrices of the projectors Π and $\Pi^{\perp}$ are easy to find if we rewrite the vectors of $\vec{x} \in \mathcal{I}_5$ in terms of the basis $\vec{u}_1, \ldots, \vec{u}_5$. (Let U be the generator matrix for this basis; then the new coordinates of $\vec{x}$ are given by the quintuple $U^{-1}(\vec{x})$.) To project into $\mathcal{E}$, we drop the last three coordinates; to project into $\mathcal{E}^{\perp}$ we drop the first two.

Now that we have found $\mathcal{E}$, the rest of the construction is straightforward. The canonical window is a rhombic icosahedron (see Figure 6.21), and we project onto $\mathcal{E}$ the set X of lattice points that lie in the cylinder with this polyhedron as 'base'. By Proposition 2.18, there are several different ways to determine which points of $\mathcal{L}^p$ lie in X.

2.7 Lattices with icosahedral point groups

The crystallographic restriction says that no point lattice in E^3 is invariant under the rotations of a regular icosahedron. Although five-fold rotation is possible for lattices in E^4, we encounter the full icosahedral symmetry for the first time when $n = 6$. (This can be proved with the help of group representation theory; see e.g. Kramer and Haase, 1989.)

2.7.1 The $\mathcal{I}_6$ quasicrystal

Consider, for example, the points $(\pm\tau, \pm 1, 0)$ and the points obtained from them by cyclic permutations of the coordinates: they are the 12 vertices of an icosahedron centered at the origin. Here, as throughout this book, $\tau = (1 + \sqrt{5})/2$ is the 'golden number'; its fascinating properties and its important role in quasicrystal theory are discussed in Chapter 4. By Hadwiger's theorem these 12 vectors form a eutactic star. Thus (after appropriate rescaling) they lift to the orthonormal cross

$$\pm\vec{e}_1 = \pm(1,0,0,0,0,0), \ldots, \pm\vec{e}_6 = \pm(0,0,0,0,0,1)$$

in E^6. The vectors $\vec{e}_1, \ldots, \vec{e}_6$ are a basis for the six-dimensional hypercubic lattice $\mathcal{I}_6$. Any permutation of the 12 vectors in E^3 induces a permutation of the corresponding vectors in E^6, and thus the point group of $\mathcal{I}_6$ must have a subgroup isomorphic to H_3, the symmetry group of the icosahedron (the notation is explained below). The group H_3 has the presentation $r^5 = s^3 = (rs)^2 = 1$, and it is easy to check that the rotations

$$R = \begin{pmatrix} 0 & 0 & 0 & 0 & 0 & 1 \\ 1 & 0 & 0 & 0 & 0 & 0 \\ 0 & 1 & 0 & 0 & 0 & 0 \\ 0 & 0 & 1 & 0 & 0 & 0 \\ 0 & 0 & 0 & 1 & 0 & 0 \\ 0 & 0 & 0 & 0 & 1 & 0 \end{pmatrix}$$

and

$$S = \begin{pmatrix} 0 & -1 & 0 & 0 & 0 & 0 \\ 0 & 0 & -1 & 0 & 0 & 0 \\ 1 & 0 & 0 & 0 & 0 & 0 \\ 0 & 0 & 0 & 0 & 0 & 1 \\ 0 & 0 & 0 & 1 & 0 & 0 \\ 0 & 0 & 0 & 0 & 1 & 0 \end{pmatrix}$$

satisfy these relations.

The group H_3 has two totally irrational, mutually orthogonal, three-dimensional stable spaces. Choosing one of them to be $\mathcal{E}$ and the other $\mathcal{E}^\perp$, it is straightforward to construct a quasicrystal by canonical projection. We will study this quasicrystal further in Chapter 7 in connection with the three-dimensional analogue of the Penrose tilings.

2.7.2 The root lattices

Although $\mathcal{I}_n$ is the lattice most frequently used in quasicrystal theory, many interesting quasicrystal point sets can be constructed with other lattices. Three of the highly symmetrical 'root' lattices have been used for this purpose.

Definition 2.12 The root lattices are the set of lattices characterized by two properties:

  (i)   their point groups are irreducible, and
  (ii) all their facet vectors are short vectors.

It is clear from the definition that the point group of a root lattice is the symmetry group of its set of short vectors.

This definition is nonstandard, but it is equivalent to the usual one: the root lattices are those whose point groups are irreducible and generated by reflections and, if necessary, the central inversion $x \to -x$ (see Senechal, 1992). The root lattices have been studied intensively, partly because the tools for studying them are highly developed, and partly because they arise in other mathematical contexts, such as sphere packings and Lie algebra theory (from whence comes the term "root"). They are discussed in detail in Conway and Sloane (1988) and in Moody and Patera (1992a). *The root lattices are integral lattices.*

A group generated by reflections is elegantly described by its Coxeter diagram, a graph whose nodes represent the generating reflections and whose edges represent the angles between the reflection hyperplanes. Nodes representing orthogonal hyperplanes are not connected by an edge, while hyperplanes that meet at an angle of π/k, where k is an integer $\neq 2$, are indicated by labeling the edge joining their nodes by k. (When $k = 3$, the label is usually omitted.) The diagrams for the symmetry groups of the three-dimensional regular solids and some of their two- and four-dimensional analogues are shown in Figure 2.14. A group generated by reflections is irreducible if and only if its diagram is connected.

The diagram contains a great deal of information about each group and also about relations between groups. For example, it is easy to deduce some group–subgroup relations directly from the diagram: if we suppress some of the nodes of a diagram, we are left with a subset of the generators, which then generate a subgroup of the group. (Look for subgroups of the groups shown in Figure 2.14.)

Some of the finite irreducible groups can be extended to discrete infinite

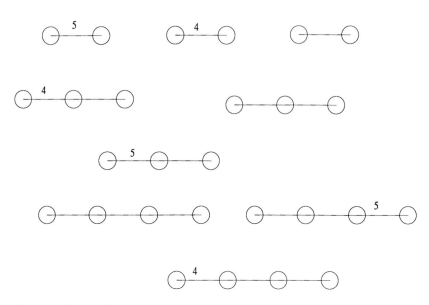

Fig. 2.14 Coxeter diagrams for the symmetry groups of the equilateral triangle, the square, the regular pentagon, the cube, the regular tetrahedron, and the regular icosahedron, and their four-dimensional analogues. Which is which?

groups by including another reflection among the generators. For example, the group whose diagram is shown in Figure 2.15(a) is generated by two reflections in lines making an angle of $\pi/3$ (Figure 2.15(b)). We can add a third reflection line in such a way that the three lines bound an equilateral triangle (Figure 2.15(c)); the extended diagram is shown in Figure 2.15(d). These infinite groups are crystallographic groups in the sense of Chapter 1; their translation lattices are the root lattices.

There are two infinite families of root lattices, denoted A_n and D_n, and one finite family, E_n, $n = 6, 7, 8$ (Figure 2.16), where n is the dimension of the lattice. Every root lattice has a basis consisting of short vectors.

The A_n and D_n lattices are sublattices of the integer lattice $\mathcal{I}_n$.

The lattice A_n is invariant under $(n + 1)$-fold rotation (note that its extended diagram is a $(n + 1)$-gon). This and other properties of the A_n lattices are most easily understood by embedding them in E^{n+1} as n-dimensional sublattices of $\mathcal{I}_{n+1}$:

$$(x_1, \ldots, x_{n+1}) \in A_n \longleftrightarrow (x_1, \ldots, x_{n+1}) \cdot (1, 1, \ldots, 1) = 0.$$

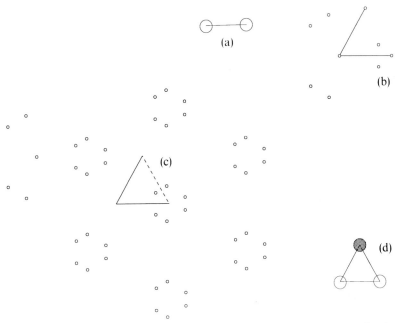

Fig. 2.15 The group whose diagram is shown in (a) is generated by two reflections in planes making an angle of $\pi/3$; (b) the points are an orbit of this finite group. (c) We can add a third reflection plane in such a way that the three planes bound an equilateral triangle; now the orbit is infinite. (d) The extended diagram: the extending node is shaded.

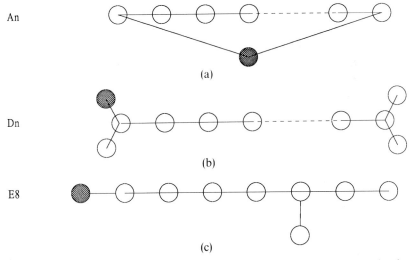

Fig. 2.16 The root lattice familes (a) A_n, (b) D_n, and (c) E_8. To obtain the diagrams for their point groups, delete the shaded node.

The short vectors all have the form $(1, -1, 0, \ldots, 0)$; that is, the entries are a 1, a -1, and $(n-1)$ 0s. There are $2\binom{n+1}{2} = n(n+1)$ vectors of this type; they are the facet vectors of A_n. Thus the Voronoï cell of A_3 has 12 facets and that of A_4 has 20. The lattice A_4 has two stable two-dimensional subspaces under the five-fold rotation; they are precisely the planes $\mathcal{E}$ and $\mathcal{E}'$ constructed in Section 2.6. Thus it is possible to construct two-dimensional quasicrystals from A_4 (see Baake, Kramer, Schlottmann, and Zeidler, 1990).

The D_n lattice is sometimes called the checkerboard lattice because it is the sublattice of $\mathcal{I}_n$ that consists of 'every other' vector. More precisely, it is defined by the condition:

$$(x_1, \ldots, x_n) \in D_n \longleftrightarrow \sum_{j=1}^{n} x_j \in 2\mathbb{Z}.$$

Its Voronoï cell has $4\binom{n}{2} = 2n(n-1)$ facets and $2^n + 2n$ vertices; when $n = 3$, $V(0)$ is a rhombic dodecahedron, and for higher dimensions it is a generalization of a rhombic dodecahedron. When $n = 4$ the vertices are all equidistant from the origin and the Voronoï cell is the regular polytope known as the 24-cell (Coxeter, 1973).

To construct the E_8 lattice, we use the fact (see e.g. Conway and Sloane, 1988) that the vertices of the Voronoï cell of D_8 belong to two different (but congruent) orbits with respect to the point group. Choose one of them, and consider *its* orbit under all the symmetries of D_8; it turns out that this orbit is a congruent copy of D_8. The lattice E_8 is the union of D_8 and the copy we have just constructed; its lattice points have coordinates

$$(x_1, \ldots, x_8), \quad \sum_{j=1}^{8} x_j \in 2\mathbb{Z} \text{ where all } x_j \in \mathbb{Z} \text{ or } \mathbb{Z} + 1/2$$

with respect to the orthonormal basis $\vec{e}_1, \ldots, \vec{e}_8$ for $\mathcal{I}_8$. The E_8 Voronoï cell has 240 facets. The other two members of the E_n family are sublattices of E_8.

2.7.3 D_6 and E_8 quasicrystals

Both D_6 and E_8 have been used to construct icosahedral quasicrystals. The first case is a straightforward application of the canonical projection method, so we describe it only briefly. The second is unusual and we will discuss it in more detail.

To construct the D_6 icosahedral quasicrystal (see, e.g., Baake, Joseph, Kramer, and Schlottmann, 1990), we first determine its two invariant orthogonal three-dimensional subspaces. We saw that, in the projections of $\mathcal{I}_5$ (and A_4), the basis vectors projected to the vertices of a regular pentagon and pentagram in $\mathcal{E}$ and $\mathcal{E}'$ respectively (Figure 2.3). In the present case, they project to the vertices of a regular icosahedron and a stellated icosahedron. The Voronoï cell for D_6 has 60 facets and 76 vertices; among these vertices are those of a hypercube $\mathcal{Q}_6$. When we project the cell into $\mathcal{E}^\perp$ the remaining vertices map into the interior of the projection of this hypercube. Thus the window for D_6 is the same as that for $\mathcal{I}_6$. However, since the two Voronoï cells have different numbers and arrangements of five-dimensional facets, the local configurations in these quasicrystals are very different (see Chapter 6).

The E_8 quasicrystal is obtained a little differently: the three-dimensional quasicrystal is a section of a four-dimensional one that is obtained by projection from the eight-dimensional lattice E_8. This construction is due to Elser and Sloane (1987).

At first glance it is not obvious that E_8 could have anything to do with icosahedra, but the relationship becomes visible if we look at the facet vectors of the E_8 Voronoï cell from a different point of view.

Recall (the standard source is Duval, 1964) that the quaternions

$$a + bi + cj + dk = (a, b, c, d)$$

form a real four-dimensional vector space (with basis $1, i, j, k$) that is equipped with a product called quaternion multiplication:

$$i^2 = j^2 = k^2 = -1, \ ij = -ji = k, \ jk = -kj = i, \ ki = -ik = j.$$

The conjugate of $q = a + bi + cj + dk$ is $\bar{q} = a - bi - cj - dk$ and the product of a quaternion and its conjugate is its *quaternionic norm:*

$$Q(q) = q\bar{q} = a^2 + b^2 + c^2 + d^2.$$

$Q(q)$ is a nonnegative real number.

Among the quaternions a particular set I with quaternionic norm 1 plays a special role. These 120 quaternions, obtained by changes of sign and even permutations from the vectors

$$(\pm 1, 0, 0, 0), \quad \frac{1}{2}(\pm 1, \pm 1, \pm 1, \pm 1), \quad \frac{1}{2}(0, \pm 1, \pm \sigma, \pm \tau)$$

where $\sigma = -1/\tau$, are called *icosians;* they form a multiplicative group.

Table 2.1. *Correspondence between I and the facet vectors of E_8 (From Elser and Sloane, 1987).*

Icosian	E_8 vector
$(1,0,0,0)$	$\vec{e}_1 + \vec{e}_2$
$(0,1,0,0)$	$\vec{e}_2 + \vec{e}_6$
$(0,0,1,0)$	$\vec{e}_3 + \vec{e}_7$
$(0,0,0,1)$	$\vec{e}_4 + \vec{e}_8$
$(\sigma,0,0,0)$	$\frac{1}{2}(-\vec{e}_1 + \vec{e}_2 + \vec{e}_3 + \vec{e}_4 + \vec{e}_5 - \vec{e}_6 - \vec{e}_7 - \vec{e}_8)$
$(0,\sigma,0,0)$	$\frac{1}{2}(-\vec{e}_1 - \vec{e}_2 + \vec{e}_3 - \vec{e}_4 + \vec{e}_5 + \vec{e}_6 - \vec{e}_7 + \vec{e}_8)$
$(0,0,\sigma,0)$	$\frac{1}{2}(-\vec{e}_1 - \vec{e}_2 - \vec{e}_3 + \vec{e}_4 + \vec{e}_5 + \vec{e}_6 + \vec{e}_7 - \vec{e}_8)$
$(0,0,0,\sigma)$	$\frac{1}{2}(-\vec{e}_1 + \vec{e}_2 - \vec{e}_3 - \vec{e}_4 + \vec{e}_5 - \vec{e}_6 + \vec{e}_7 + \vec{e}_8)$

There is also a *Euclidean norm* for the icosians: if $Q(q) = x + y\sqrt{5}$, then

$$N(q) = x + y.$$

If an icosian has quaternionic norm 1, it also has Euclidean norm 1, but the converse is not true. There are 240 icosians of Euclidean norm 1, and indeed there is a one to one correspondence between them and the facet vectors of E_8. The correspondence is specified in Table 2.1.

It can be shown that I is invariant under the reflection group H_4 (the four-dimensional analogue of H_3). It follows that the Voronoï cell of the E_8 lattice is also invariant under this group. The group stabilizes two four-dimensional orthogonal subspaces, both of which are totally irrational.

Thus we have all the requirements for canonical projection. The Voronoï cell of E_8 projects to a four-dimensional polytope in $\mathcal{E}^\perp$; projection onto $\mathcal{E}$ gives us a *four-dimensional* icosahedral quasicrystal. To obtain a three-dimensional one, we take a three-dimensional section (for instructions, see Elser and Sloane (1987)). The local structure of this quasicrystal can be determined in a straightforward way, but evidently no one has done it, nor have connections with tilings been explored.

It is easy to see, by deleting the shaded nodes in Figure 2.16, that the point groups satisfy $A_4 \subset D_6 \subset E_8$. It is not so obvious that A_4, D_6, and E_8 contain subgroups isomorphic to H_2, H_3, and H_4 respectively. However, even this information can be extracted from the diagrams in a very elegant way (Scherbak, 1988). Thus these quasicrystals belong to a single family.

2.7.4 Is the projection formalism necessary?

'The six independent vectors of the reciprocal space required in the analysis of diffraction patterns of three-dimensional quasicrystals displaying icosahedral symmetry are no longer a matter of contention among physicists', note Moody and Patera (1993). 'More contentious is whether or not this fact demands a hyperspace theory formulated in higher than three-dimensional spaces . . . we develop a theory that allows the three- and six-dimensional worlds to live together simultaneously in the same space; it is only a matter of interpretation which of the two is being discussed.'

The Moody–Patera theory, in which the root lattice quasicrystals described in the preceding sections appear together in a unified way, shows that, at least for these quasicrystals, the projection formalism can be thought of as a language rather than an architecture.

We will give only a bare outline of some aspects of the theory here; even so, the tools needed take us a bit beyond the scope of this book.

First, the set-up. Let $F = Q + Q\sqrt{5}$ be the extension of the rational numbers by $\sqrt{5}$. Conjugation, the map

$$\prime : F \to F, \quad (a + b\sqrt{5})' = a - b\sqrt{5},$$

is an automorphism of F. The ring of integers of F is the ring $Z[\tau] = Z + Z\tau$.

Let H denote the quaternionic algebra over R and let

$$H_F = F + Fi + Fj + Fk.$$

On H_F we set

$$(a, b, c, d)' = (a', b', c', d').$$

Now recall the correspondence between icosians and the facet vectors of E_8 (Table 2.1). Let S be the set of facet vectors of E_8, V their rational linear combinations, and V_F their linear combinations with coefficients in F. Let II be the Z-span of the icosians; it is closed under complex conjugation $(\overline{(a, b, c, d)} = (a, -b, -c, -d))$ but not under $\prime$.

Next, we construct 'projections' that can be identified with the usual orthogonal projections. Set

$$\Pi : S \to I \cup \tau I, \quad \Pi^{\perp} : S \to I' \cup \sigma I'.$$

We can also construct a linear map $T : V \to V$ 'which mimics in V the multiplication of icosians by τ' by setting

$$\Pi(Tx) = \tau\Pi x.$$

T is called the inflation map of V (relative to Π). We can extend T, Π and $\Pi^\perp$ to V_F; V_F then splits into two four-dimensional stable subspaces under T.

Now we define a 'quasicrystal', or rather a family of 'quasicrystals' parametrized by the positive real numbers: they are the sets

$$\sum^r = \{x \in \Pi | Q(x') < r^2\},$$

where Q is the quaternionic norm. Moody and Patera show that $\sum^r$ is discrete and invariant under H_4. No reference to a higher-dimensional structure is needed, but one can show that

$$\sum^r = \{\Pi(x) | x \in Q, \Pi^\perp(x) \cdot \Pi^\perp(x) < r^2\}.$$

Thus it can also be thought of as a projected set, with a spherical window. We use quotation marks above because it has not been proved that these Delone sets are crystals according to Definition 1.1. (If they are crystals, then they are certainly quasi.) By restricting to pure quaternions ($x = -\bar{x}$) we obtain H_3-invariant three-dimensional Delone sets, and restricting further to appropriate complex planes we obtain H_2-invariant two-dimensional Delone sets.

A remarkable property of these point sets is that each of them is closed under the operators T_x, for every $x \in \sum^r$, defined by

$$T_x y = \tau^2 x - \tau y.$$

It is easy to see that $T_x x = x$, while all other points of $\sum^r$ are inverted in x and dilated by the factor τ. Interestingly, some of the sets are generated by a finite number of such operators; thus we do not need the higher-dimensional formalism.

This construction lends weight to Mermin's contention (Mermin, 1991) that 'superspace' is not needed for the classification of quasicrystals (though it may be a useful way to study local configurations and other properties).

2.8 Notes

1. For an introduction to mathematical crystallography, see Senechal (1991b).

2. The pioneers of n-dimensional crystallography were Hermann and Schwarzenberger; Schwarzenberger's book on the subject (1980) was inspired by the work of Hermann. The crystallographic restriction for dimensions greater than three was formulated by Hermann in (1949).

3. To maximize symmetry, crystallographers sometimes use unit cells of volume larger than $|\det \mathcal{L}|$, containing more than one lattice point; they are called 'centered cells' in the crystallographic literature. However, this strategem does not always solve the problem. For example, no parallelepiped of any size has hexagonal symmetry.

4. Voronoï cells are also known as Dirichlet domains, in honor of the mathematician who first (about 1850) introduced them into mathematics in order to clarify some concepts in number theory. However, the construction is evidently much older – Descartes used a version of it in 1644 to describe the structure of the heavens! (See Descartes, 1644 or Okabe, Boots, and Sugihara, 1992.) The same construction has subsequently been redisovered or reintroduced in many different contexts and consequently is known by many other names, including "Wigner-Seitz cells", and "Brillouin zones". Voronoï extended the concept to higher-dimensional lattices, which is why we choose to use his name for the cells. Readers interested in the many generalizations and applications of Voronoï diagrams should consult Okabe, Boots, and Sugihara.

5. A famous conjecture states that every convex polytope that fills E^n by translation is affinely equivalent to a Voronoï polytope; the conjecture has been proved for $n \leq 4$.

6. The canonical projection method is due to de Bruijn (1981a), who used it to derive the fundamental properties of Penrose tilings (see Chapter 6); we will use it throughout this book. The construction is closely related to the 'models' for harmonious sets discussed by Meyer (1972). Kramer and Neri (1984) appear to have been the first to suggest its usefulness for crystallography. The first point set models for quasicrystals were obtained by canonical projection from $\mathcal{I}_6$ (see e.g. Katz and Duneau, 1986; Kalugin, Kitaev, and Levitov, 1986; Levine and Steinhardt, 1986).

7. Is there a quasicrystallographic restriction? Although we can construct projected Delone sets with (local) rotational symmetry of any order, all of the real quasicrystals discovered so far have either icosahedral symmetry or rotational symmetry of orders 8, 10, or 12.

8. No discussion of quasicrystal symmetry is complete that does not include its dilation properties (thus our discussion is incomplete). The interested reader should consult Janner (1991).

3

Introduction to diffraction geometry

The effects produced can be more symmetrical than their causes.

(Pierre Curie)

Introduction

Crystals have always been studied through their images – even crystal form is an image, in the same sense that a forest is an image produced by its constituent trees. Since 1912 the x-ray diffraction pattern has been the principal image through which crystals are studied; today x-ray images are supplemented by electron and neutron diffraction images, and images produced by HREM, STM, and other high-technology techniques.

Diffraction is another name for the scattering that takes place when waves meet an obstacle. The scattered waves spread out, 'interfere', and recombine; if sufficiently many recombined waves are in phase, then distinct images (auditory, visual) can be reconstructed by an appropriate receiver (for example, your ear or your eye, unaided or aided by technology).

X-rays are 'light' waves, analogous to audible and visible waves, but on an extremely small, subvisible, scale. Fortunately, the scale is exactly right for the atomic pattern of a crystal to serve as a diffraction grating. Although we cannot see these diffraction images directly, the intensities of the diffracted waves can be recorded on photographic plates (or, more commonly today, measured by computer). When there are a great many scatterers, arranged in such a way that the scattered waves are in phase (or nearly so) in some directions, then the combined intensities are very large in those directions and appear in the photographs as sharp, bright spots ('Bragg peaks').

Optical patterns produced by visible light diffracted by a plane grating or *mask* are two-dimensional analogues of the x-ray diffraction images of crystals; this analogy is sometimes used in crystallography to explore the rudiments of diffraction geometry (Lipson and Lipson, 1981). We will follow this custom. Figure 3.1, from the visual dictionary *Atlas of Optical*

Fig. 3.1 Masks (left) and diffraction patterns (right). (From *Atlas of Optical Transforms, Harburn, Taylor, and Welberry, 1975.*)

Transforms (Harburn, Taylor and Welberry, 1975), shows several particularly striking examples of masks and diffraction patterns.

In introductory optics, one distinguishes between near-field or Fresnel diffraction, and far-field or Fraunhofer diffraction. Fresnel diffraction is rather complicated mathematically, but Fraunhofer diffraction is easily modeled by the elegant tool of Fourier transformation. The optical diffraction patterns in Figure 3.1 are Fraunhofer patterns, as are x-ray diffraction patterns produced by crystals.

Although the relation between mask and pattern may seem mysterious when you first look at Figure 3.1, it can easily be understood once the alphabet of this language has been deduced. Not only can we reproduce these patterns with the help of simple computer programs (Figure 3.2), we

Fig. 3.2 Computer simulation of one of the diffraction patterns in Figure 3.1.

can also compute the patterns that would be produced by masks that we design ourselves.

We will begin our discussion at the beginning, with the basic concepts of diffraction geometry. Our goal is to develop a mathematical model for diffraction when the mask or diffraction 'grating' is a general Delone set Λ. In the process of constructing the model from first principles we encounter some interesting mathematical questions.

3.2 Diffraction geometry

Fraunhofer diffraction, as the term 'far-field' suggests, is characterized by the property that the distances from the light source to the mask and from the mask to the diffraction screen or photographic plate are 'far'. More precisely, these distances are assumed to be sufficiently large to permit us to make two enormously simplifying assumptions (Figure 3.3):

• light waves from a point source arrive at all points of the mask in parallel trains, and

• parallel trains of diffracted light making the same angle with the mask meet at the same point on the diffraction screen.

With the additional assumption that the light is monochromatic – that there is a single wavelength λ – we can compute the patterns created by one, two, and many point scatterers. (See Section 3.5 for more details about the assumptions underlying these computations.)

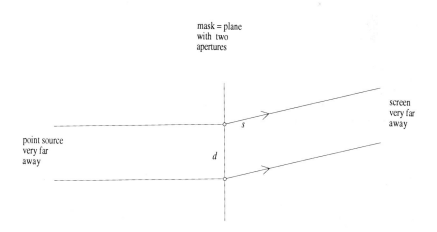

Fig. 3.3 Model for Fraunhofer diffraction.

3.2.1 One-point diffraction

Figure 3.4 shows a sequence of patterns produced by masks with a single circular aperture. The corresponding diffraction patterns are also circular: they are illuminated disks (surrounded by concentric rings). Notice that the radius of the diffraction disk is inversely proportional to the radius of the aperture.

Evidently, as the radius of the aperture decreases to zero, the radius of the diffraction disk increases without bound. We may suppose, then, that 'in the limit' –if the aperture were a point of zero diameter – the entire (infinite) screen would be illuminated with equal intensity.

We will take this 'observation' as an axiom. We assume that the aperture is a single point, and we let $J(\vec{s})$ be the value (on the screen or photographic plate) of the intensity of the wave front in the direction of the (unit) vector $\vec{s}$. $J(\vec{s})$ has the same value in all directions $\vec{s}$; our axiom is:

Axiom A $J(\vec{s}) \equiv 1$.

Proposition 3.1 The value of $J(\vec{s})$ is independent of the position of the aperture in the diffraction mask.

We know (from physics) that $J(\vec{s})$ is the squared modulus of the amplitude function $A(\vec{s})$:

$$J(\vec{s}) = |A(\vec{s})|^2. \tag{3.1}$$

The amplitudes of the diffracted waves are not constant in all directions and do vary with the position of the aperture.

It is not difficult to determine just how $A(\vec{s})$ must vary: (3.1) and Axiom A together imply that $A(\vec{s})$ must be a complex-valued function of modulus 1. This means, in exponential notation,

$$A(\vec{s}) = \exp(-2\pi i f(\vec{s})), \tag{3.2}$$

where f is a real-valued function of s. In other words, $A(\vec{s})$ is a wave function. By considering diffraction by two apertures, we can determine the function $f(\vec{s})$.

3.2.2 Two-point diffraction

We now assume that the mask has two point apertures, located at points P_1 and P_2; let $\vec{d}$ be the vector from P_1 to P_2. Now the diffracted waves will interfere and the intensity will no longer be constant. Instead, $J(\vec{s})$ will vary with $\vec{s}$ in a way that depends on $\vec{d}$.

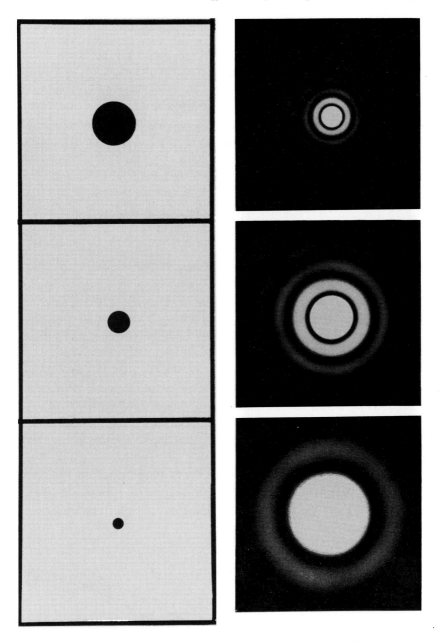

Fig. 3.4 As the radius of the aperture decreases, the radius of the diffraction spot increases. (From *Atlas of Optical Transforms.*)

Light waves emanating from a point source will be in phase when they reach the mask, but they may not be in phase when they reach the screen. As Figure 3.5 shows, one front may lag behind the other.

The length σ of the lag is the projection of $\vec{d}$ on $\vec{s}$; since $|\vec{s}| = 1$ we have

$$\sigma = \vec{d} \cdot \vec{s} = d \cos \theta, \tag{3.3}$$

where $d = |\vec{d}|$ and θ is the angle between $\vec{s}$ and $\vec{d}$.

We can draw some useful conclusions from Equation (3.3). We express the lag in units of the wavelength λ:

$$\sigma = u\lambda,$$

where u is some real number. Then the projection of $\vec{s}$ on $\vec{d}$ is

$$\vec{d} \cdot \vec{s} = u\lambda = \cos \theta. \tag{3.4}$$

Equation (3.4) does not specify $\vec{s}$ uniquely: for each value of u, there is an infinite set of vectors satisfying this condition. Since all the vectors of the set make the same angle with $\vec{d}$, they lie on a cone, and since they all have the same length, they project to a single line (Figure 3.6) on a diffraction screen parallel to $\vec{d}$. The lines, which are orthogonal to $\vec{d}$, can thus be labeled by the real numbers.

If σ is an integral number n of wavelengths λ, then the wave fronts emerging from the two apertures will arrive at the screen in phase; thus directions for which $\sigma = n\lambda$, where n is an integer, correspond to lines of maximal intensity. The amplitudes *cancel* when $\sigma = n\lambda/2$ and n is odd; along these lines the intensity is zero. Thus the diffraction pattern will consist of a continuous infinite family of parallel lines that vary periodically from light to dark; the period is proportional to $1/d$. (The wavelength λ will not play any further role in our discussion, so we assume that it is fixed once and for all, and suppress it in our notation.)

Fig. 3.5 One front lags behind the other.

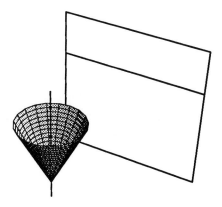

Fig. 3.6 Directions $\vec{s}$ such that $\vec{s} \cdot \vec{d} = u\lambda$ form a cone and project to a line on the diffraction screen.

Returning to $A(\vec{s})$, we note that since wave amplitudes are additive, $A(\vec{s})$ is the sum

$$A(\vec{s}) = A_0(\vec{s}) + A_d(\vec{s}). \tag{3.5}$$

We know, by Proposition 3.1, that

$$|A_0(\vec{s})|^2 = |A_d(\vec{s})|^2 = 1, \tag{3.6}$$

so we must have

$$A_d(\vec{s}) = z_d A_0(\vec{s}) \tag{3.7}$$

for some complex number $z_d = z_d(\vec{s})$, with $|z_d| = 1$. Since z_d can be written in exponential form and the multiplication of two exponentials is additive in their arguments, we see again that *translation of the aperture effects a phase shift in the amplitude*. This phase shift is the lag $\vec{d} \cdot \vec{s}$, so multiplication by z_d transforms $A_0(\vec{s})$ in a way that varies inversely with $\vec{d}$:

$$A_d(\vec{s}) = \exp(2\pi i\vec{d} \cdot \vec{s}) A_0(\vec{s}). \tag{3.8}$$

Substituting (3.8) in (3.5) we have

$$A(\vec{s}) = A_0(\vec{s})\bigl(1 + \exp(2\pi i\vec{d} \cdot \vec{s})\bigr). \tag{3.9}$$

Now it is straightforward to compute the intensity:

$$\begin{aligned} J(\vec{s}) &= A(\vec{s})\overline{A(\vec{s})} \\ &= |A_0(\vec{s})|^2\bigl(1 + \exp(2\pi i\vec{d} \cdot \vec{s})\bigr)\bigl(1 + \exp(-2\pi i\vec{d} \cdot \vec{s})\bigr) \\ &= 2\bigl(1 + \cos(2\pi\vec{d} \cdot \vec{s})\bigr). \end{aligned} \tag{3.10}$$

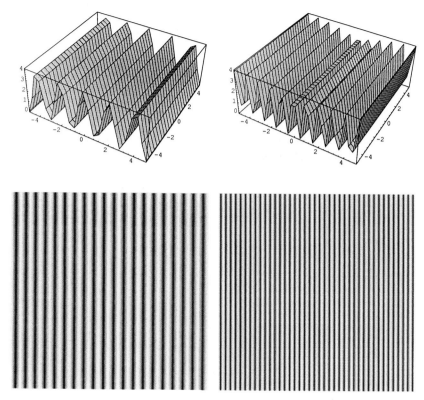

Fig. 3.7 Top: graphs of intensity functions for two-point diffraction. Bottom: the spacing of the maxima in the corresponding diffraction pattern.

The intensity $J(\vec{s})$ attains its maxima when $\cos(2\pi\vec{d}\cdot\vec{s}) = 1$; that is, it attains its maxima on the lines (on the screen) corresponding to $\vec{d}\cdot\vec{s} \in Z$. It attains its minima when $\cos(2\pi\vec{d}\cdot\vec{s}) = -1$; these lines are midway between the lines of maxima. The spacing of the maxima is inversely proportional to d.

Figure 3.7 is the one-picture 'alphabet' of the language of diffraction geometry. Every diffraction pattern produced by a mask of point apertures, no matter how the points are arranged, is a superposition of families of dark and light bands, one family for each pair of points. Since amplitudes are additive, points lying in the intersection of many lines of maximum brightness will be especially bright. If the points are not colinear, then the families of parallel lines will intersect, producing complex patterns of light and darkness.

3.2.3 N-point diffraction

Figure 3.8 shows the intensity function and diffraction patterns of masks with three, four, and five point apertures located at the vertices of regular polygons. Here, and from now on, we identify the unit vector $\vec{s}$ with the point s at which the line containing it intersects the diffraction screen; the origin of the plane of the screen is identified with the unit vector orthogonal to the mask. With these identifications, we can also think of $\vec{s}$ as the vector $s - 0$ and $J(\vec{s})$ as a function of two variables. The 'diffraction patterns' are gray-scale maps of the square roots of the intensity functions, produced by the computer program 'fourier' whose code is given in Chapter 8.

The computation of the intensity function for any finite number of points is based on (3.10). For $N + 1$ point apertures at $d_0 = 0, d_1, \ldots, d_N$, we have

$$A(\vec{s}) = A_0(\vec{s}) + A_1(\vec{s}) + \cdots + A_N(\vec{s})$$

$$= A_0(\vec{s}) \left(1 + \sum_{j=1}^{N} \exp(2\pi i \vec{d}_j \cdot \vec{s})\right) \tag{3.11}$$

and

$$J(\vec{s}) = |A(\vec{s})|^2 = \sum_{j=0}^{N} \sum_{k=0}^{N} \exp(2\pi i (\vec{d}_j - \vec{d}_k) \cdot \vec{s})$$

$$= N + 1 + 2 \sum_{j=1}^{N} \sum_{k<j} \cos(2\pi(\vec{d}_j - \vec{d}_k) \cdot \vec{s}). \tag{3.12}$$

Notice that $J(\vec{s}) = J(-\vec{s})$: this is the reason why diffraction patterns are centrosymmetric. Notice also that the value of the sum on the right-hand side depends only on the *relative* positions of the points in the mask.

If the arrangement is sufficiently 'regular', then the maxima of the intensity increase as $(N + 1)^2$. Let us consider an example where the points are colinear. Figure 3.9 shows the graphs of $J(\vec{s})$ for three sets of $N + 1$ equally spaced points, $N = 5, 8, 13$. (Here we have computed a cross-section of $J(\vec{s})$, since the intensity function is constant in the direction orthogonal to the line on which the points lie.) Neither the number nor the spacing of the peaks changes with N; the effect of increasing N is to make the peaks higher and narrower *and the light parts of the diffraction pattern sharper and brighter.*

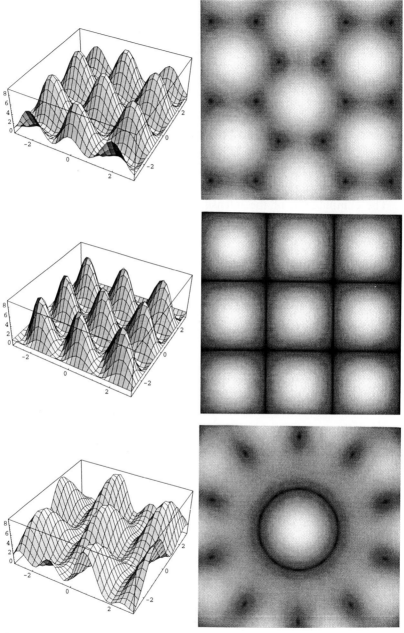

Fig. 3.8 All diffraction patterns are superpositions of families of parallel lines, as in Figure 3.7. Here the point apertures of the masks are – from top to bottom – located at the vertices of an equilateral triangle, a square, and a regular pentagon. Left: the graphs of the intensity functions; right: the computer-simulated diffraction patterns (gray-scale maps).

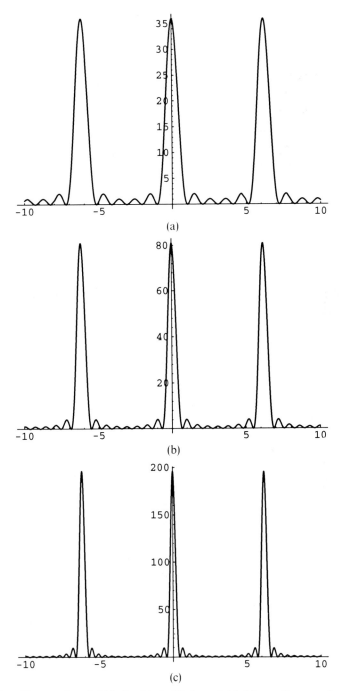

Fig. 3.9 The graphs of $J(\vec{s})$ for (a) 6, (b) 9, and (c) 14 equally spaced colinear points.

If we modify Figure 3.9 by changing the arrangement of points so that some of the relative distances between them are irrational, the graph of $J(\vec{s})$ changes dramatically. For example, Figure 3.10 also shows the intensity functions for six, nine, and fourteen points on a line, but now the distances between successive points are not equal: they assume two values, and the ratio between 'short' and 'long' intervals is $\tau : 1$ where, as usual, $\tau = (1 + \sqrt{5})/2$. Notice that the local maxima are not much smaller than the maxima in the periodic case: what we see here is a simple finite approximation to the quasicrystal phenomenon. We will study this example extensively in Chapter 4.

We conclude this section with three instructive diffraction patterns (Figure 3.11). Each was produced by a mask with approximately 500 point apertures. The mask for the top picture is a square region of the standard lattice $\mathcal{I}_2$. The apertures of the mask for the middle pattern also lie at the points of a square region of $\mathcal{I}_2$, but in this case approximately half the sites are vacant (the choice was made randomly). The mask for the bottom picture has no lattice structure: the aperture positions, which also lie in a square, were chosen by a random process. As we look down the page, we see clearly a progressive deterioration of 'order'.

When there is a countable infinity of apertures – the case that more closely models crystal structure – the sums (3.11) and (3.12) are infinite. This raises problems of interpretation and computation.

3.3 Fourier transforms and the Wiener diagram

There has been a dissymmetry in our discussion so far – the diffraction pattern has been described by a function, $J(\vec{s})$, while the the mask of apertures – a set of points – is a geometric object. This is unsatisfactory: we need an analytical expression for the scatterers, too.

In optics textbooks it is shown that in diffraction of the Fraunhofer type, $A(\vec{s})$ is the *Fourier transform* of a 'function' $\rho(\vec{x})$ describing the arrangment and nature of the scatterers (see e.g. Lipson and Lipson, 1981; Cowley, 1981). The function ρ is sometimes called a transmission function or, in the case of diffraction by real crystals, the *density function*. We have used quotation marks above because when the apertures (or atoms) are points, ρ is not a function as functions are usually defined; it is a *generalized function*. But whatever $\rho(\vec{x})$ is, it must be the inverse Fourier transform of $A(\vec{s})$ – in some sense.

Thus Fourier transformation plays a central role in diffraction theory: in the words of Lipson and Lipson, 'the whole of Fraunhofer diffraction

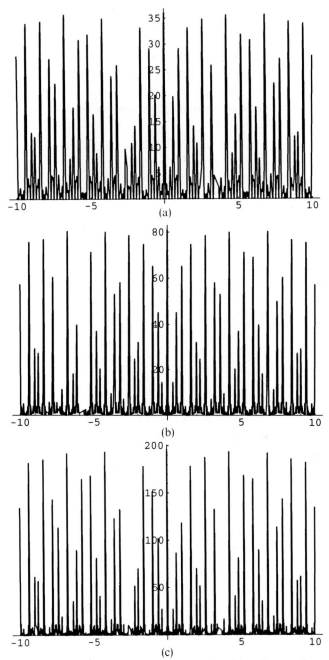

Fig. 3.10 The graphs of $J(s)$ for (a) 6, (b) 9, and (c) 14 colinear points with two spacings of relative sizes.

Quasicrystals and geometry

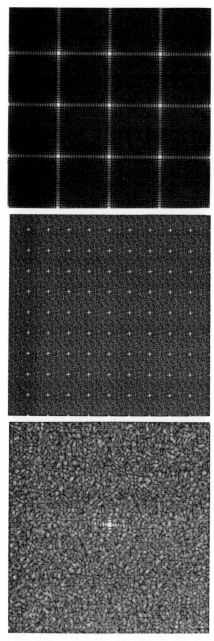

Fig. 3.11 The diffraction patterns produced by three masks, each with approximately 500 point apertures. From top to bottom: a square region of $\mathcal{I}_2$, a randomly occupied square region of $\mathcal{I}_2$, and a randomly chosen set of points lying in a square.

theory can be summed up in one statement: the Fraunhofer diffraction pattern of an object is the Fourier transform of that object'. While this statement is not exactly true (aside from its dissymmetry, it is the amplitude $A(\vec{s})$, not the intensity $J(\vec{s})$, that is the Fourier transform of the density), it does not overstate the importance of this mathematically delicate tool.

In this section we outline the most useful Fourier relations among ρ, A, and J. We begin with a brief review of some important facts about Fourier transformation for 'ordinary' functions (see e.g. Chandrasekharan, 1989 or Champeney, 1987).

We first note that not all functions have Fourier transforms in the 'classical' sense. The Fourier transform of a function f of n variables is the function $\hat{f}$ defined by the integral

$$\hat{f}(\vec{s}) = \int_{R^n} f(\vec{x}) \exp(-2\pi i \vec{x} \cdot \vec{s}) \, d\vec{x}; \tag{3.13}$$

if the integral does not converge (for almost all $\vec{s}$) then the transform cannot exist. For example, if $f(\vec{x}) \equiv 1$, the integral does not exist for any $\vec{s}$. On the other hand if f is the characteristic function of a bounded, measurable set C, that is, if

$$f(\vec{x}) = \begin{cases} 1, & \text{if } \vec{x} \in C, \\ 0, & \text{otherwise.} \end{cases}$$

then the transform does exist.

Some basic properties of the classical transform are listed without proof in Theorem 3.1. We assume that the functions f and g are complex-valued functions on E^n and belong to a linear space S of functions which is closed under Fourier transformation. All of these properties can be extended to generalized functions.

Recall the distinction between Fourier transformation and the Fourier transform: Fourier transformation is an operator (to be denoted by $\mathcal{F}$) on the space S that associates to a function f its Fourier transform $\hat{f}$:

$$\mathcal{F}f = \hat{f}. \tag{3.14}$$

Theorem 3.1 $\mathcal{F}$ has the following properties (among many others):
   (i) $\mathcal{F}$ is linear:

$$\mathcal{F}(f + g) = \mathcal{F}f + \mathcal{F}g \tag{3.15}$$

and

$$\mathcal{F}af = a\mathcal{F}f, \tag{3.16}$$

where a is a complex scalar.

(ii) $\mathcal{F}$ is invertible: there is an operator $\mathcal{F}^{-1}$ such that for all x (except possibly a set of measure zero):

$$f(\vec{x}) = \mathcal{F}^{-1}\hat{f}(\vec{s}) = \int_{R^n} \hat{f}(\vec{s})\exp(2\pi i \vec{s}\cdot\vec{x})d\vec{s}. \tag{3.17}$$

Note the change of sign in the exponential function. We say that $\hat{f}$ and f are Fourier pairs.

(iii) If f is translated, then $\hat{f}$ is rotated (just as in (3.8)), and vice versa:

$$\mathcal{F}f(\vec{x}+\vec{c}) = \exp(-2\pi i \vec{c}\cdot\vec{x})\hat{f}(\vec{x}). \tag{3.18}$$

(iv) If A is a linear isometry, then

$$\mathcal{F}f(A\vec{x}) = \hat{f}(A\vec{s}). \tag{3.19}$$

(v) The Fourier transform of the complex conjugate of $f(-\vec{x})$ is the complex conjugate of $\hat{f}(\vec{s})$:

$$\mathcal{F}(\overline{f(-\vec{x})}) = \overline{\hat{f}(\vec{s})}.$$

(vi) The Fourier transform of a convolution product is the ordinary product of the Fourier transforms and vice versa:

$$\mathcal{F}(f * g) = \mathcal{F}(f)\mathcal{F}(g) \quad \text{and} \quad \mathcal{F}(fg) = \mathcal{F}(f) * \mathcal{F}(g). \tag{3.20}$$

When $g(\vec{x}) = \overline{f(-\vec{x})}$, the convolution product is called the *autocorrelation function of* f (recall that $f * g(\vec{x}) = \int_{R^n} f(\vec{t})g(\vec{x}-\vec{t})\,d\vec{t}$). By (v) and (vi), its Fourier transform is the squared modulus of $\hat{f}$. Thus we have

Corollary 3.1 The following diagram commutes:

$$
\begin{array}{ccc}
f(\vec{x}) & \overset{convolution}{\longrightarrow} & f(\vec{x}) * \overline{f(-\vec{x})} \\
\updownarrow & & \updownarrow \\
\hat{f}(\vec{s}) & \overset{squaring}{\longrightarrow} & |\hat{f}(\vec{s})|^2
\end{array} \tag{3.21}
$$

Setting $f(\vec{x}) = \rho(\vec{x})$ and $\gamma(\vec{x}) = f(\vec{x}) * \overline{f(-\vec{x})}$, we can see how relevant this diagram would be to our diffraction model if the operations were valid for the objects involved:

$$\rho(\vec{x}) \qquad \longrightarrow \qquad \gamma(\vec{x})$$

$$\updownarrow \qquad\qquad\qquad \updownarrow \qquad\qquad (3.22)$$

$$A(\vec{s}) = \hat{\rho}(\vec{s}) \quad \longrightarrow \quad J(\vec{s}) = \hat{\gamma}(\vec{s})$$

This diagram shows, in a simple way, the relations among the key functions of x-ray crystallography. The central difficulty for the x-ray crystallographer is that the observed $J(\vec{s})$ does not contain all the information needed to determine $\rho(\vec{x})$: information about the phases has been lost in the passage to $J(\vec{s})$. But to determine the structure of the crystal a route from $J(\vec{s})$ to $\rho(\vec{x})$ must be found. Diagonal routes are not possible; one must try to go by either a west–north route, or a north–west one. The former requires restoring the lost information (this can often be handled by 'direct methods', a technique that earned the Nobel Prize in Chemistry for its developers, Herbert Hauptman and Jerome Karle). The latter involves 'deconvolution' of the autocorrelation function (this is known as Patterson analysis). To paraphrase Lipson and Lipson, *the whole of x-ray diffraction theory is summed up in this one diagram!* Unfortunately its validity is problematical, for reasons explained below. But it is still useful to keep the diagram in mind.

We will call (3.22) a *Wiener diagram* in honor of Norbert Wiener who, in the early 1930s, pointed out the usefulness of the autocorrelation function for deciphering x-ray diffraction patterns (Patterson, 1962). In addition, Wiener's pioneering work in generalized harmonic analysis (Wiener, 1930) is central to a deeper understanding of diffraction geometry.

3.4 Dirac deltas and Dirac combs

Let us return to the problem of finding the scattering function $\rho(\vec{x})$ explicitly, where ρ is the density for a set of $N+1$ point apertures at positions $x_0, \ldots, x_N$ in the mask. This will be the inverse Fourier transform of $A(\vec{s})$; by (3.11) and Theorem 3.1,

$$\rho(\vec{x}) = \mathcal{F}^{-1}\left(A_0(\vec{s})\left(1 + \sum_{j=1}^{N} \exp(-2\pi i \vec{x}_j \cdot \vec{s})\right)\right)$$

$$= \mathcal{F}^{-1}\left(A_0(\vec{s})\right) * \left(\mathcal{F}^{-1}(1) + \sum_{j=1}^{N} \mathcal{F}^{-1}\left(\exp(-2\pi i \vec{x}_j \cdot \vec{s})\right)\right);$$

the factor on the right is

$$\int_{R^n} 1 \exp(2\pi i \vec{x} \cdot \vec{s})\, d\vec{s} + \sum_{j=1}^{N} \int_{R^n} \exp(-2\pi i(\vec{x} - \vec{x}_j) \cdot \vec{s})\, d\vec{s}. \qquad (3.23)$$

Using (3.8) we can easily find all of these inverse transforms if we know the inverse transform of any one of them, say the constant function 1.

For simplicity, let us consider functions of one variable. None of the integrals in (3.23) converges, but we can give them meaning by passing to generalized functions. Let $f(s) \equiv 1$ and replace f by a truncated version of itself, the characteristic function of the interval $[-T, T]$. Since this function is zero outside the interval, its inverse Fourier transform, $\rho_T(x)$, is given by

$$\int_{-T}^{T} 1 \exp(2\pi i x s)\, ds. \qquad (3.24)$$

For finite values of T, the integral exists and can be evaluated:

$$\rho_T(x) = \frac{\sin(2\pi T x)}{\pi x}. \qquad (3.25)$$

Note that if we let $u = Tx$, the right-hand side of (3.25) becomes

$$T \frac{\sin(2\pi u)}{\pi u}.$$

The inverse Fourier transform of f should be the limit

$$\rho(x) = \lim_{T \to \infty} \frac{\sin(2\pi T x)}{\pi x}. \qquad (3.26)$$

The graphs of $\rho_T(x)$ for $T = 2, 10$, and 50 are shown in Figure 3.12. We see that the functions become increasingly narrow and increasingly steep. The limit of this sequence is the famous Dirac delta. That is,

$$\mathcal{F}^{-1}(1) = \delta. \qquad (3.27)$$

The functions $\rho_T(x)$ belong to a large family of functions of a single parameter that 'converge' to the Dirac delta. The delta is characterized by the property

$$\int_{-\infty}^{\infty} \delta(x) g(x)\, dx = g(0) \qquad (3.28)$$

for sufficiently well-behaved functions g. Equation (3.27) is a special case:

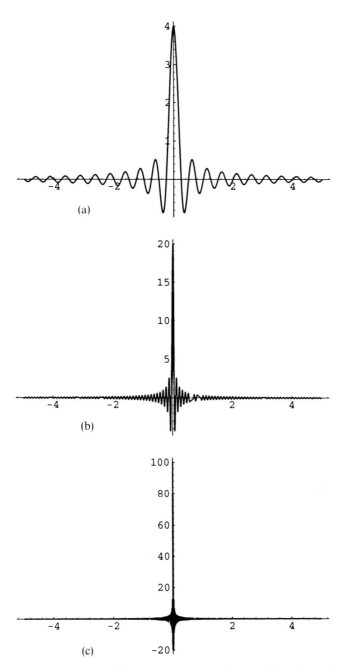

Fig. 3.12 The functions $\rho_T(x)$ belong to a large family of functions of a single parameter that 'converge' to the famous Dirac delta, $\delta(x)$; here $T = 2, 10, 50$.

when $g(x) = \exp(-2\pi i x s)$ we have

$$\int_{-\infty}^{\infty} \delta(x) \exp(-2\pi i x s)\, dx = 1. \tag{3.29}$$

There are many different theories of generalized functions; the best known are those of Schwartz (1952) and Lighthill (1958). (Another, due to de Bruijn (1973), is outlined in Appendix II.) In Schwartz's theory of tempered distributions, generalized functions are linear functionals and the Dirac delta is the particular linear functional that associates the function $g(x)$ to the real number $g(0)$ (this is the content of (3.28)). Lighthill's generalized functions are the limits of (classes of) sequences of functions; the delta is the limit of various sequences, including (3.25). Schwartz's and Lighthill's classes of generalized functions can be shown to be equivalent, but in other theories the generalized functions are defined by subtly different properties; for example, de Bruijn's theory is not equivalent to these (van Eijndhoven, 1987). But in every theory an appropriately generalized version of Theorem 3.1 holds; in particular it is always the case that the space of generalized functions is closed under Fourier transformation, and

Proposition 3.2 $\delta(x - d)$ and $\exp(-2\pi i d s)$ are Fourier pairs.

Moreover, the Dirac delta (together with all of its remarkable properties) can always be generalized to spaces of arbitrary dimension. Spaces of generalized functions are linear spaces, so finite sums of Dirac deltas are also generalized functions (as are certain infinite sums, under suitable definitions of convergence). Thus the inverse Fourier transform of (3.5) is $\delta(\vec{x}) + \delta(\vec{x} - \vec{d})$. More generally, the following generalized functions are Fourier pairs:

$$\rho(\vec{x}) = \sum_{d_k \in \Lambda} \delta(\vec{x} - \vec{d}_k), \tag{3.30}$$

$$\hat{\rho}(\vec{s}) = \sum_{d_k \in \Lambda} \exp(-2\pi i \vec{d}_k \cdot \vec{s}) \tag{3.31}$$

for any Delone set Λ (these sums converge as generalized functions; see Appendix II).

De Bruijn observed that when Λ is a one-dimensional point lattice, $\rho(x)$ must look something like a comb with infinitely long teeth (see Figure 3.9); he called it a *Dirac comb*. We will use this terminology for any countably infinite sum of (possibly weighted) deltas. The set of points Γ at which the deltas are located need not be discrete.

Definition 3.1 A Dirac comb is any sum of the form

$$\sum_{d_k \in \Gamma} c(k)\delta(\vec{x} - \vec{d_k}),$$

where the $c(k)$s are complex numbers and the sum converges (as a generalized function).

Is the Wiener diagram valid for Dirac combs? This is a complicated question, because in general the operations of convolution and multiplication are not defined for all generalized functions (in any theory). However, at least in some theories there is a subclass of functions – that includes finite sums of deltas – for which one of these operations can be made meaningful, and it turns out nicely that if two functions can be convoluted then their Fourier transforms can be multiplied, and vice versa. It follows that the diagram is valid for the densities of *finite* point sets. Only very special Dirac combs belong to the convolution subclass, however, and the ones we are interested in are not among them (see Appendix II). This issue is also discussed in Hof (1993) in the context of Schwartz distributions. It is possible that the diagram can be justified in some more general context (see e.g. Ruelle, 1970).

3.5 The diffraction condition

The Dirac delta plays two indispensable roles in diffraction theory. The first, which we have just discussed, is that of place marker for the points of a Delone set Λ. Deltas may also appear in the diffraction patterns themselves.

Let Λ be a Delone set in E^n, with points $d_k, k \in Z$. Then

$$\rho_\Lambda(\vec{x}) = \sum_{d_k \in \Lambda} \delta(\vec{x} - \vec{d_k})$$

is a Dirac comb. What does its diffraction pattern look like? Will it be, in the terminology of Definition 1.1, 'essentially discrete', that is, will it show sharp, bright spots? If so then, according to Definition 1.1, Λ is a crystal. Or will it instead be 'essentially continuous' – hazy with (or without) regions of relatively bright intensity maxima? The two top pictures in Figure 3.11 illustrate the first possibility, and the bottom illustrates the second. The answer depends, of course, on the positions of the points in Λ or, what amounts to the same thing, on the behavior of the Fourier transform $\hat{\gamma}$ of its autocorrelation function γ. Setting $\Lambda_T = \Lambda \cap [-T, T]^n$, we can define

$$\hat{\gamma}(\vec{s}) = \lim_{T \to \infty} \frac{1}{(2T)^n} \mid \sum_{d_k \in \Lambda_T} \exp(-2\pi i \vec{d}_k \cdot \vec{s}) \mid^2. \qquad (3.32)$$

(For the technical reasons for this formulation of $\hat{\gamma}$, see Hof, 1993.)

Definition 3.2 Let Λ be a Delone set in E^n with density ρ and autocorrelation γ. The spectrum of Λ is its 'intensity function' $\hat{\gamma}$, defined by (3.32).

It can be shown that $\hat{\gamma}(\vec{s})$ is a positive measure and thus, by a well-known theorem of Lebesgue, we can decompose it uniquely into a sum of two components, one discrete and the other continuous:

$$\hat{\gamma}(\vec{s}) = \sum_{s^* \in S} c(\vec{s}) \delta(\vec{s} - \vec{s}^*) + \hat{\gamma}_c(\vec{s}); \qquad (3.33)$$

the coefficients $c(s)$ are real numbers. The set S is always countable; when it is countably infinite, the first term on the right, the discrete component of the intensity, is a Dirac comb. (In this context, 'discrete' refers to the measure, not to S, which may be dense in the diffraction screen.) The continuous component of the spectrum may itself be a sum of 'singular continuous' and 'absolutely continuous' components. We always have $0 \in S$; if $S = \{0\}$, $\hat{\gamma}$ is said to be purely continuous, while if $\hat{\gamma}_c \equiv 0$ it is said to be purely discrete. When the spectrum is purely discrete, $\hat{\gamma}$ is itself a Dirac comb.

We will say that a Delone set *diffracts*, or satisfies the diffraction condition, if there is a countable infinity of deltas in its spectrum; that is, if the set S in (3.33) is countably infinite. We do not require $\hat{\gamma}_c \equiv 0$.

Definition 3.3 (the diffraction condition) Λ is a crystal if its spectrum has a countably infinite discrete component.

Throughout this book we will encounter many nonperiodic but crystalline Delone sets, and also several examples of Delone sets with continuous spectra.

Now we can sharpen Definition 1.2:

Definition 3.4 The symmetry group of Λ is the group G of isometries that leave its intensity function invariant:

$$\phi \in G \quad \longleftrightarrow \quad \hat{\gamma}(\phi \vec{s}) = \hat{\gamma}(\vec{s}). \qquad (3.34)$$

3.6 Searching for deltas

In trying to determine which Delone sets Λ satisfy the diffraction condition, we encounter two serious obstacles. The first is the doubtful validity of the Wiener diagram. The second problem is that even if the diagram is valid, determining the decomposition (3.33) can be a formidable problem.

If the diagram were valid, then life would be simpler for us. As we will see in later chapters, there are powerful techniques for showing that $\hat{\rho}(s)$ is a Dirac comb, or at least has a component that is a Dirac comb. With the diagram, we could conclude that the Delone set in question is a crystal. Without it, we can only conclude that every (suitably chosen) finite approximation to the Delone set is an approximation to a crystal.

Let us agree to accept the diagram as a 'working concept', in order to get on with our story, and to hope that it will eventually be justified. In any case, Hof (1993) has shown that, for sets obtained by the projection method, the weights of the discrete part of $\hat{\gamma}$ are in fact the squares of the weights of $\hat{\rho}$, so using the diagram will not lead us grievously astray.

What we are seeking, intuitively, are directions $\vec{s}^{*}$ for which all the expressions $\vec{d}_{k} \cdot \vec{s}^{*}$ that appear in (3.31) are identical modulo 1 (or nearly so). For these directions, letting $z_{d} = \exp(-2\pi i \vec{d}_{k} \cdot \vec{s}^{*})$ we have, approximately and formally,

$$\hat{\rho}(\vec{s}^{*}) = z_{d} \sum_{d_{k} \in \Lambda} 1. \tag{3.35}$$

It follows that there will be bright spots at these $\vec{s}^{*}$ in the diffraction pattern.

There is always at least one such point s^{*} for every Delone set Λ: $s^{*} = 0$. This is why the origin of a diffraction pattern is always a spot of maximal brightness. *If Λ is a point lattice $\mathcal{L}^{p}$ or a subset of $\mathcal{L}^{p}$, then every point s^{*} of the dual lattice is also a point of maximal brightness, since $\vec{d}_{k} \cdot \vec{s}^{*} \in Z$.* In fact, when $\Lambda = \mathcal{L}^{p}$ we have $S = \mathcal{L}^{p*}$ and $\hat{\gamma}_{c}(\vec{s}) \equiv 0$; this is the content of one form of the famous *Poisson summation formula*:

Theorem 3.2 (Poisson summation formula) Let $\mathcal{L}$ and $\mathcal{L}^{*}$ be dual n-dimensional lattices. Then

$$\mathcal{F}\left(\sum_{\ell \in \mathcal{L}} \delta(\vec{x} - \vec{\ell})\right) = \sum_{\vec{\ell}^{*} \in \mathcal{L}^{*}} \delta(\vec{s} - \vec{\ell}^{*}). \tag{3.36}$$

This powerful formula (for a proof, see Appendix II) is one of the most

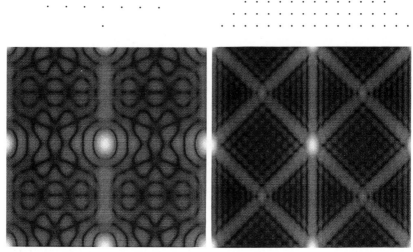

Fig. 3.13 The finite size effect: a circular and a triangular region of a square point lattice are shown, together with their diffraction patterns. (See also the square region in Figure 3.11(top).)

useful in all of mathematical crystallography, both classical and quasi. It says that the diffraction pattern of a point lattice is the dual point lattice. It should be noted, however, that it is a statement about infinite sums of generalized functions, and is true only in a generalized framework. For finite sums there are regions of varying intensity; this is known as the 'finite size effect' (Figure 3.13). For example, if we think of the density $\rho_t(\vec{x})$ of the triangular point set in Figure 3.13 as the product of the density $\rho(\vec{x})$ of the lattice and the characteristic function χ of the triangle, then we should have

$$\hat{\rho}_t(\vec{s}) = \hat{\rho}(\vec{s}) * \hat{\chi}(\vec{s});$$

the factor $\hat{\chi}(\vec{s})$, which contributes the 'varying intensities', is an echo of the shape of the finite region.

3.6.1 Lattice-vacancy point sets

As a simple but very interesting type of nonperiodic crystal, let us consider infinite subsets of point lattices. Let $\mathcal{L}^p$ be a point lattice in E^n, and Λ be any subset of $\mathcal{L}^p$. The density ρ associated to Λ has the form

$$\rho(\vec{x}) = \sum_{\ell \in \Lambda} \delta(\vec{x} - \vec{\ell}). \tag{3.37}$$

By the definition of the dual lattice $\mathcal{L}^*$, we have

$$\vec{\ell} \cdot \vec{\ell}^* \in Z, \quad \text{for every } \vec{\ell}^* \in L^*$$

and so

$$\hat{\rho}(\vec{s}) = \sum_{\ell \in \Lambda} \exp(-2\pi i \vec{\ell} \cdot \vec{s}) = \sum_{\ell \in \Lambda} 1$$

for every $\vec{s} \in \mathcal{L}^*$. Since $\mathcal{L}^*$ is countably infinite, Λ is a crystal. This is why we see bright spots in the middle pattern of Figure 3.11, even though the vacancies in the lattice were randomly chosen.

3.6.2 Poisson combs

When both ρ and $\hat{\rho}$ are Dirac combs, then $\hat{\gamma}(\vec{s}) \equiv 0$ and

$$\mathcal{F}\Big(\sum_{d_k \in \Lambda} \delta(\vec{x} - \vec{d}_k)\Big) = \sum_{\vec{s}^* \in S} c(\vec{s}) \delta(\vec{s} - \vec{s}^*). \tag{3.38}$$

This equation looks so much like the Poisson summation formula (3.36) that we will call ρ and $\hat{\rho}$ *Poisson combs*. Unless Λ is a point lattice, S will always be dense in the ambient space (Cordoba, 1988).

To show that nonperiodic Poisson combs exist, we will construct a large family of them. These combs are one-dimensional, 'periodically modulated' point lattices (Bombieri and Taylor, 1985). (Other classes of Poisson comb will be constructed in later chapters.) Poisson combs are always crystals (thanks to the Wiener diagram).

To keep technicalities to a minimum, we will assume that the sums we deal with converge in the context of generalized functions and that all of the operations that we will perform on them are valid.

Let Λ be the set of points on the real line defined by

$$x_n = \alpha n + g(n), \ n \in Z, \tag{3.39}$$

where α is a fixed real number, and $g(x)$ is a periodic function of period ω, an irrational number; we assume that α and g are such that Λ is a Delone set.

Setting

$$\rho(x) = \sum_{n \in Z} \delta(x - x_n), \qquad (3.40)$$

we have

$$\hat{\rho}(s) = \sum_{n \in Z} \exp(-2\pi i(\alpha n + g(n))s). \qquad (3.41)$$

Now $\exp(-2\pi i g(x)s)$ is also periodic in x, with period ω, so we can expand it in a Fourier series:

$$\exp(-2\pi i g(x)s) = \sum_{m \in Z} c_m(s) \exp(2\pi i \frac{mx}{\omega}), \qquad (3.42)$$

where $c_m(s)$ is the mth Fourier coefficient.

Thus

$$\hat{\rho}(s) = \sum_{m \in Z} c_m(s) \left(\sum_{n \in Z} \exp(-2\pi i n(\alpha s + \frac{m}{\omega})) \right).$$

Using the Poisson summation formula to rewrite the inner sum we obtain

$$\hat{\rho}(s) = \sum_{m \in Z} \sum_{n \in Z} c_m(s) \delta(\alpha s - \frac{m}{\omega} - n).$$

Since

$$\delta(\alpha s - \frac{m}{\omega} - n) = \delta(s - (\frac{m}{\omega} + n)/\alpha),$$

we have

$$\hat{\rho}(s) = \sum_{m \in Z} \sum_{n \in Z} c_m((\frac{m}{\omega} + n)/\alpha) \delta(s - (\frac{m}{\omega} + n)/\alpha). \qquad (3.43)$$

Thus $\hat{\rho}(s)$ is a weighted sum of deltas and Λ is a Poisson comb!

This completes a sketch of the proof of

Theorem 3.3 (Bombieri and Taylor) The Delone set Λ defined by (3.39) is a Poisson comb.

Notice that, because ω is irrational, the real numbers $(\frac{m}{\omega} - n)/\alpha$ are dense in the real line. This means that the bright spots should be everywhere dense in the diffraction pattern. This is not what we see, however: some bright spots are brighter than others and some cannot be seen at all, due to the weights c_m.

Theorem 3.3 is one of several general methods that establish the diffraction condition for large classes of Dirac combs by showing that they are in fact Poisson combs; we will describe other methods in the next chapter. But Λ need not be a Poisson comb to satisfy the diffraction condition, since (3.33) does not require that $\hat{\gamma}(\vec{s}) \equiv 0$. Any technique that determines whether $\hat{\rho}$ (or $\hat{\gamma}$) has a nontrivial discrete component is sufficient for our purposes.

Equation (3.35) suggests that we can try to 'detect' the presence of deltas in a Fourier transform by searching for vectors s for which $\hat{\rho}$ increase like N as we sum over an appropriately chosen nested sequence of finite subsets

$$\Lambda_0 \subseteq \Lambda_1 \subseteq \Lambda_2 \subseteq \dots, \text{ where } \cup \Lambda_i = \Lambda. \tag{3.44}$$

If we can show that the limit

$$\lim_{N \to \infty} \frac{1}{|\Lambda_N|} \hat{\rho}_N(\vec{s}) \tag{3.45}$$

exists and is nonzero, then the diffraction pattern will have a point of maximal brightness at $\vec{s}$; this means that $\hat{\gamma}(\vec{s})$ has a delta at $\vec{s}$ (see Hof (1993), Theorem 3.4).

3.7 Notes

1. The idea of a language of diffraction patterns is discussed in historical context by Glaser and Wrinch (1953). Our discussion is rather nonstandard; in particular I have never seen Axiom A anywhere in the literature. Its chief virtue is that it enables us to skip over a great deal of physics and go straight to the geometry of diffraction patterns.

2. I am grateful to Professor Mark Heald, Professor Emeritus of Physics, Swarthmore College, for the following comments.

(i) In the context of Fraunhofer diffraction, 'very far away' is a distance greater than D^2/λ, where D is the maximum extent of the mask's aperture pattern. This is equivalent to saying that the spherical wavefront diverging from the source point (or converging on the observation point) is negligibly different from a plane wavefront over the aperture, within the tolerance of a small fraction of a wavelength.

(ii) Apertures of nonzero size, such as those shown in Figure 3.7, are analyzed as superpositions of an infinite number of infinitesimal sources. This is why the diffraction pattern of a discrete set of apertures, all of the same nonzero size, is a convolution of the diffraction pattern one would obtain were the apertures simply points with the 'shape factor' of the aperture itself.

(iii) That we have straight lines, not parabolas, in Figures 3.6 and 3.7 is due to two 'paraxial approximations'. One is the assumption that subtends a small angle at the observation point, and the other that the dimension of the observable diffraction pat   1 subtends small angle at the mask. The former ($r^2 >> D^2$) is subsumed under the Fraunhofer limit ($r >> D^2/\lambda$) and the assumption of a radiation field ($r >> \lambda$), where r is the distance between the mask and the observation plane. The latter is independent of the Fraunhofer assumption, and requires $D^2 >> \lambda^2$, which is not necessarily true in the present context, but usually is.

(iv) The reader should be aware that there are various conventions about where to put the 2π factor in the Fourier transform; there are also different conventions regarding sign. Moreover, physicists usually use overbars to signify averages.

3. The Dirac delta, $\delta(x)$ is often introduced as the 'function' with the properties

$$\delta(x) = \begin{cases} 0, & \text{if } x \neq 0, \\ \infty, & \text{otherwise,} \end{cases} \qquad (3.46)$$

together with the normalizing condition

$$\int_R \delta(x)dx = 1. \qquad (3.47)$$

But since there is no function with these properties, this definition is unacceptable.

The notation $\delta(\vec{x})$ is also unacceptable to many mathematicians because it suggests that the Dirac δ is a function of a variable x in the ordinary sense (writing $\hat{\gamma}(\vec{s})$ is, I realize, even worse). However, efforts to abolish this notation lead to other complications, so we will retain it despite its misleading suggestivity.

4. The word 'spectrum' is used in many ways in mathematics, not all of them equivalent or even closely related, even when we are speaking of the spectrum of functions. For example, according to Wiener, the spectrum of a function f is an *antiderivative* of the Fourier transform of its autocorrelation. In tilings associated with dynamical systems or sub-

stitutions, the spectrum of the translation operator on the space of all tilings (of the appropriate type) is studied. Dworkin (1993) has shown that this latter spectrum is consistent with the diffraction spectrum we study here. For a more detailed discussion of the notion of spectrum for tiling dynamical systems, see Solomyak (1993).

5. Among many open problems in diffraction geometry, the following seem especially noteworthly.

(i) It would be useful to understand the algebra of diffracting sets. For example, is the diffraction property preserved under union, intersection, linear transformations, and other operations? Which subsets of diffracting sets also diffract? Is there an analogue, for the diffraction condition, of Delone's local criterion for regularity (Chapter 1)? This question will come up again in connection with the diffraction patterns in Chapter 8.

(ii) Further research is also needed on the robustness of the diffraction condition under perturbations of various kinds, such as local rearrangements of the points of Λ. Some interesting work on this problem is discussed in Chapters 4 and 5.

(iii) We need a rigorous theory of diffraction geometry, formulated within a theory of generalized functions. As de Bruijn has pointed out (1986b), 'the first thing that one should require from the analytical point of view, is that when working with generalized functions (like infinite sums of delta functions), one should make it clear to what function space these objects belong. Also when dealing with infinite processes, it has to be made clear what kind of convergence procedures are intended.' De Bruijn, and also Porter, have given rigorous proofs of the fact that canonically projected crystals have no continuous components in their spectra. But in order to *explore* the diffraction properties of point sets, especially sets that are not produced by deterministic methods such as projection or substitution, we need a rigorous theory of what we might call 'delta heuristics'. Such a theory would involve both the notion of a delta as a limit of a sequence of ordinary functions and numerical methods for the estimation of the growth of exponential sums.

(iv) Another important problem is that of correlating the concepts of one theory of generalized functions with those of another. One question is just how de Bruijn's theory differs from that of Schwarz and what the implications of these differences are. Another is the following. As Lighthill (1958) and others have shown, it is possible to construct a rigorous theory of generalized functions as limits of sequences of ordinary functions that is completely equivalent to the Schwartz theory. It would be very desirable to have such an alternative version of de Bruijn's generalized functions. Van Eijndhoven (1987) has shown that the Dirac combs of de Bruijn's space are abstractly equivalent to the generalized almost periodic functions of Burkhill and Rennie (1983), which *are* defined as limits of sequences of ordinary functions. (Specifically, he showed that both spaces can be identified with the Gelfand – Shilov spaces of type $S(1/2, 1/2)$.) But no dictionary is provided, and a dictionary is what we need in order to work with both characterizations simultaneously.

(v) Although the work of Bohr on almost periodic functions (see Chapter 1) has often been invoked to justify the use of the projection method, no analogue of the 'diagonal' theorem seems to have been proved in any theory of generalized functions. Such a theorem would provide the philosophical if not the geometrical rationale for the projection method as a central tool in crystal geometry.

6. Diffraction geometry is closely related to number theory, particularly the theory of Diophantine approximation. These connections will become more apparent in later chapters, although we will not emphasize them. Here we note that the diffraction condition is actually a statement about the asymptotic behavior of the infinite sum of exponentials

$$\hat{\rho}(\vec{s}) = \sum_{k \in \Lambda} \exp(-2\pi i \vec{k} \cdot \vec{s}) \ .$$

This is a difficult problem in number theory, even in the one-dimensional case. A series of the form

$$\sum_{n=1}^{\infty} a_n \exp(-\lambda_n s),$$

where λ_n is an increasing sequence of real numbers and $s = \sigma + it$ is complex, is called a *generalized Dirichlet series*. The series converges in the half-plane $\sigma > \sigma_o$, where

$$\sigma_o = \lim \sup \frac{\ln |\sum_1^n a_n|}{\lambda_n},$$

and it diverges for $\sigma < \sigma_o$. The line $s = \sigma_o$ is called the line of convergence;

depending on the coefficients a_n and the sequence λ_n, σ_o may be $-\infty, \infty$ or any real number.

In the special case we are interested in, $a_n = 1$ for all n and $\lambda_n = O(n)$ (because Λ is relatively dense in the real line). Then $\sigma_o = 0$ and we see that $\hat{\rho}$ is nothing other than a generalized Dirichlet series restricted to its line of convergence. And, as G. H. Hardy once noted: 'On the line of convergence the question of the convergence of the series remains open, and requires considerations of a much more delicate character.'

7. Two point sets in the same local isomorphism class will have the same diffraction pattern; Mermin (1991) calls such sets *indistinguishable* and argues that indistinguishability should be the starting point for the derivation of the symmetry groups of periodic and aperiodic crystals alike.

4

Order on the line

I wanted to persuade myself that these were only apparent
flaws, that they were all part of a much vaster regular structure,
in which every asymmetry we thought we observed really
corresponded to a network of symmetries so complicated we
couldn't comprehend it . . .

(Italo Calvino)

4.1 One-dimensional crystals

There are real materials, both periodic and nonperiodic, that can be
profitably studied as if they were one-dimensional (see e.g. (Dharma-
wardana *et al.*, 1987)). The one-dimensional sets are also rich in
mathematical interest because there are many different ways of generating
and analyzing them. In this chapter we consider one-dimensional Delone
sets and their diffraction properties. While they do not have the visual
appeal of some of their higher-dimensional counterparts (particularly
tilings), this is one of their advantages: we are not distracted by inessential
features. In particular, the one-dimensional case suggests that symmetry
may be a relatively unimportant feature of aperiodic crystals.

A one-dimensional Delone set Λ is a set of colinear points, and so can
be identified with a two-way infinite sequence of real numbers

$$\ldots, x_{-2}, x_{-1}, x_0, x_1, x_2, \ldots \tag{4.1}$$

whose successive differences or step sizes $x_{k+1} - x_k$ are bounded above
and below. Any sequence will do, so long as there exist fixed positive
numbers R_0 and r_0 such that

$$0 < r_0 \le x_{k+1} - x_k \le R_0 \tag{4.2}$$

for all $k \in Z$.

We think of the points $x_k \in \Lambda$ as point scatterers, and we want to know
if Λ is a crystal. To find out, we give Λ a density function of the form

$$\rho(x) = \sum_{x_k \in \Lambda} \delta(x - x_k) \tag{4.3}$$

and compute its Fourier transform

$$\hat{\rho}(s) = \sum_{x_k \in \Lambda} \exp(-2\pi i x_k s). \tag{4.4}$$

The question is whether the intensity $\hat{\gamma}(s)$ satisfies the diffraction condition (Definition 3.3). In some cases the answer will be yes, in others no; the problem is to find the characteristics that distinguish the two cases. This general problem remains unsolved. In this chapter we discuss some special cases which have received a great deal of attention.

In Chapter 3 we showed that one-dimensional Delone sets whose points deviate from a point lattice by a periodic modulation are always crystals; indeed their diffraction spectra are purely discrete. In that family of sets, although there are upper and lower bounds for the step sizes, there is no restriction on the *number* of different step sizes. Now we will restrict ourselves to the case where the number of step sizes is finite. Then Hadwiger's theorem (2.2) allows us to reinterpret Λ as a projection of a path in the standard n-dimensional point lattice $\mathcal{I}_n^p$. The geometry of the path provides important information about the behavior of the sum (4.4).

The paths we will construct are called *staircases* because they are ascending and do not loop back. However, they are not well suited for climbing because of their apparent unpredictability.

Here is the general scheme. Finiteness of the number of steps means that for all $k \in Z$,

$$x_{k+1} - x_k \in \{\alpha_1, \ldots, \alpha_n\}, \tag{4.5}$$

where the step sizes α_j are positive real numbers. Setting $x_0 = 0$, the coordinate of the mth point of Λ is

$$x_m = m_1 \alpha_1 + \cdots + m_n \alpha_n$$
$$= (m_1, \ldots, m_n) \cdot (\alpha_1, \ldots, \alpha_n) \tag{4.6}$$

where m_j is the number of occurrences of α_j in the first m steps, and $\sum_{j=1}^n m_j = m$. Let

$$\vec{\alpha} = (\alpha_1, \ldots, \alpha_n) \in E^n,$$

rescaled (if necessary) to length one. Let $\vec{e}_1, \vec{e}_2, \cdots, \vec{e}_n$ be an orthonormal basis for E^n such that $\vec{e}_j \cdot \vec{\alpha} = \alpha_j$, $j = 1, \ldots, n$. The lattice generated by $\vec{e}_1, \vec{e}_2, \cdots, \vec{e}_n$ is the standard lattice $\mathcal{I}_n$. To construct the staircase, we concatenate the vectors $\vec{e}_j$ as prescribed by the sequence x_k (Figure 4.1).

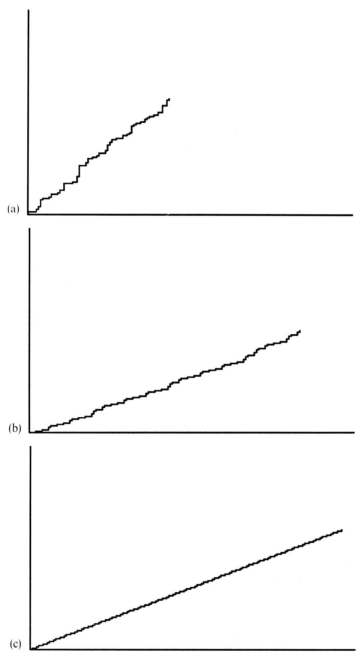

Fig. 4.1 The staircases corresponding to three nonperiodic sequences with two step sizes: (a) a randomly generated sequence, (b) an LB sequence, (c) an F sequence (LB and F sequences are defined later in this chapter).

We will call the vectors that occur in (4.6),

$$\vec{p}_m = (m_1, \ldots, m_n), \quad \sum_{j=1}^{n} m_j = m,$$

population vectors for the Delone set Λ. We can recover Λ by projecting the nodes of the staircase (that is, the lattice points that lie on it) onto the line generated by $\vec{\alpha}$.

Questions we might ask about such a staircase include:
- Does it have an 'average slope'?
- If so, do the population vectors stay close to it?
- What do the answers to the previous question tell us about the diffraction condition?

We will keep these questions in mind as we discuss several different methods for constructing Delone sets and for computing the Fourier transforms of their densities. We introduce each of these methods with the famous Fibonacci crystal, one of the few that can be constructed by all of them.

First, in the next section we review Fibonacci numbers and introduce Fibonacci strings and Fibonacci sequences.

4.2 Fibonacci numbers, strings, and sequences

You are probably familiar with the Fibonacci *numbers F_n,*

$$1, \ 1, \ 2, \ 3, \ 5, \ 8, \ 13, \ 21, \ 34, \ldots, \tag{4.7}$$

defined by the recursion

$$F_{n+1} = F_n + F_{n-1}, \tag{4.8}$$

with initial conditions $F_0 = 0, F_1 = 1$.

Leonardo of Pisa (Fibonacci) introduced these numbers in 1202 as the successive sizes of a rabbit population consisting of pairs of babies (B) and pairs of adults (A). The population grows rapidly: it takes only one month for each baby pair B to grow into an adult pair A, and for each A pair to produce a new B pair. All adult rabbits are always fertile and no rabbit ever dies, so the the population will increase from one B pair to 233 A and B pairs of rabbits in just one year!

For this and other reasons, it is advisable to forget about the rabbits and think of A and B as abstract symbols. Writing the growth rule in additive notation:

$$A \longrightarrow A + B, \quad B \longrightarrow A, \tag{4.9}$$

we see that the rule is a linear transformation with matrix

$$\mathcal{P} = \begin{pmatrix} 1 & 1 \\ 1 & 0 \end{pmatrix}. \tag{4.10}$$

An initial population of one B pair corresponds to an initial population vector

$$\vec{p}_0 = (0, 1). \tag{4.11}$$

Then in the first generation we have

$$(0, 1) \begin{pmatrix} 1 & 1 \\ 1 & 0 \end{pmatrix} = (1, 0) \quad \text{(one } A\text{)}, \tag{4.12}$$

in the second we have

$$(1, 0) \begin{pmatrix} 1 & 1 \\ 1 & 0 \end{pmatrix} = (0, 1) \begin{pmatrix} 1 & 1 \\ 1 & 0 \end{pmatrix}^2 = (1, 1) \quad \text{(one } A \text{ and one } B\text{)}, \tag{4.13}$$

and so on, so that there are F_m A pairs and F_{m-1} B pairs in the $(m + 1)$st generation. Thus

$$\vec{p}_0 \mathcal{P}^{m+1} = (0, 1) \begin{pmatrix} 1 & 1 \\ 1 & 0 \end{pmatrix}^{m+1} = (0, 1) \begin{pmatrix} F_{m+1} & F_m \\ F_m & F_{m-1} \end{pmatrix} \tag{4.14}$$

$$= (F_m, F_{m-1}) = \vec{p}_{m+1}.$$

The Fibonacci numbers F_m are among the most famous constructions of elementary mathematics. Their many fascinating properties can all be traced to the matrix (4.10). Over the centuries people have looked for – and thus found – these numbers almost everywhere, just as they find fractals (or Elvis Presley) today. It is only fitting that Fibonacci numbers should be inextricably linked to that other ubiquitous mathematical phenomenon, the golden ratio:

Proposition 4.1 The ratio F_m/F_{m-1} approaches the 'golden' number τ as m increases without bound:

$$\lim_{m \to \infty} \frac{F_m}{F_{m-1}} = \frac{1 + \sqrt{5}}{2} = \tau. \tag{4.15}$$

Proof The characteristic equation of the matrix $\mathcal{P}$ is

$$\lambda^2 - \lambda - 1 = 0; \tag{4.16}$$

its solutions, the eigenvalues of $\mathcal{P}$, are τ and $-1/\tau$. The corresponding eigenvectors are multiples of the unit vectors

$$\vec{\epsilon}_1 = \frac{1}{\nu}(\tau, 1), \quad \vec{\epsilon}_2 = \frac{1}{\nu}(-1, \tau),$$

where $\nu^2 = 1 + \tau^2$. Since $\vec{\epsilon}_1$ and $\vec{\epsilon}_2$ are an orthonormal basis for E^2, we can write the population vectors and the matrix $\mathcal{P}$ in terms of them. For example, (4.11) becomes

$$\vec{p}_0 = \frac{1}{\nu}\vec{\epsilon}_1 + \frac{\tau}{\nu}\vec{\epsilon}_2 \tag{4.17}$$

and

$$\vec{p}_0\mathcal{P} = \frac{\tau}{\nu}\vec{\epsilon}_1 - \frac{1}{\nu}\vec{\epsilon}_2. \tag{4.18}$$

This shows that applying $\mathcal{P}$ to $\vec{p}_0$ has the effect of stretching its $\vec{\epsilon}_1$ component and shrinking (and flipping) its $\vec{\epsilon}_2$ component by the factor of the corresponding eigenvalue.

To show that the ratio of As to Bs approaches τ as $m \to \infty$, we write

$$\vec{p}_m = \vec{p}_0\mathcal{P}^m = \frac{\tau^m}{\nu}\vec{\epsilon}_1 + \left(\frac{(-\tau^{-1})^{m-1}}{\nu}\right)\vec{\epsilon}_2. \tag{4.19}$$

As $m \to \infty$, $(\tau^{-1})^{m-1} \to 0$ and

$$\vec{p}_m - \frac{\tau^m}{\nu}\vec{\epsilon}_1 \to 0. \tag{4.20}$$

Thus the ratio of As to Bs tends to the ratio of the components of $\vec{\epsilon}_1$, $\tau : 1$.
❑

Many useful identities can be derived from (4.16), including:

Proposition 4.2 The 'golden number' τ satisfies the following identities:

(i) $\tau^2 = \tau + 1$
(ii) $\tau - \tau^{-1} = 1$
(iii) $\tau^3 - 2\tau^2 = -1$.

The proofs are left as easy exercises.

Now let us rewrite the growth rule (4.9) in multiplicative notation:

$$A \longrightarrow AB, \quad B \longrightarrow A \tag{4.21}$$

and interpret it as a *substitution rule*: we pass from one generation of symbols to the next by substituting AB for each A and A for each B. If we follow this rule literally (that is, if we don't alter the order of the symbols) we obtain successive symbolic strings:

$$
\begin{aligned}
&B \\
&A \\
&AB \\
&ABA \\
&ABAAB \\
&ABAABABA \\
&ABAABABAABAAB \\
&ABAABABAABAABABAABABA
\end{aligned}
$$

$$\vdots$$

(4.22)

Notice that the list of symbols in each generation is a concatenation of the two preceeding ones. Notice also that although there is no obvious pattern in the sequence of As and Bs in any generation, it is clear that there are many 'forbidden words'. For example, we will never encounter BB or AAA (why not?).

As we iterate the substitution process *ad infinitum*, the length of the string of As and Bs grows without bounds.

The infinite string is nonperiodic; this follows directly from Proposition 4.1 since an infinite string of symbols (from any finite set) in which the ratio of the numbers of occurrences of some pair of symbols is irrational cannot be periodic. (A periodic string is made up of identical blocks of a fixed number of each symbol, and the ratios in the infinite sequence are those of the blocks.) However, rational ratios do not necessarily imply periodicity: there are nonperiodic sequences in which As and Bs occur with equal frequency! (The Thue–Morse sequence discussed in Section 4.7 is one of them.)

Note that (4.22) describes a string that is infinite in only one direction, and we want two-way infinite strings. There are various techniques for extending it in both directions: see e.g. (de Bruijn, 1989a; Bombieri and Taylor, 1987). But we can also define Fibonacci strings – and other strings of symbols generated by substitution rules – in a way that avoids this issue.

Consider *any* two-way infinite string of As and Bs in which the words AAA and BB never occur. We can *compose* it by grouping together the letters of each word AB and replacing them by an A (which we write as A' here, to avoid confusion), and also replacing each A (that is followed by an A, not a B) by a B':

$$A' = AB, \quad B' = A. \tag{4.23}$$

For example, this is how we can move *up* the list of finite strings in (4.22). The resulting (infinite) string is again composed of As and Bs; following de Bruijn (1981b), we call it the *predecessor* of the one we started with. This procedure is the reverse of substitution; it is called *composition*. Formally, both composition and substitution are described by the matrix $\mathcal{P}$.

If the words AAA and BB do not occur in the predecessor, we can compose again to obtain a predecessor of level two, and so on. An infinite string of As and Bs is said to have predecessors of all levels if composition can be iterated *ad infinitum* without encountering these forbidden words.

Definition 4.1 A Fibonacci string F is a two-way infinite string of As and Bs that has predecessors of all levels with respect to the composition rule (4.23).

Two infinite strings are 'different' if they cannot be superimposed by shifting one of them to the right or to the left through a finite number of symbols. Definition 4.1 admits uncountably many different Fibonacci strings (a proof can be modeled on that of Theorem 6.3).

Recalling Definition 1.7, we can state

Proposition 4.3 Every Fibonacci string is repetitive.

This is true for any point set generated by substitution, as we will prove in a more general setting in Chapter 5. It follows that all Fibonacci strings belong to a single local isomorphism class: any finite string in one is relatively dense in all of the others.

We can associate a Fibonacci string F with a Delone set by identifying A and B with line segments of two lengths, α and β. We choose the lengths to be the relative frequencies of A and B in the string: $\alpha/\beta = \tau$. The points of a Fibonacci Delone set are the endpoints of the sequence of line segments associated to the letters of F. To distinguish this point set from the string of symbols we define:

Definition 4.2 A Fibonacci sequence is a two-way infinite sequence $\ldots, -x_2, -x_1, x_0, x_1, x_2, \ldots$ of points on the real line such that

(i) $x_{n+1} - x_n \in c\{1, \tau\}$ for some $c > 0$, and
(ii) the difference sequence $x_{n+1} - x_n$ has predecessors of all levels under the composition 4.23.

After composition, the size of the large steps is $\alpha + \beta$ and the size of the small steps is α. However, their ratio is the same as before since

$$\frac{\alpha}{\beta} - \frac{\alpha + \beta}{\alpha} = \frac{\alpha^2 - \alpha\beta - \beta^2}{\alpha\beta} = \frac{\tau^2 - \tau - 1}{\tau} = 0. \qquad (4.24)$$

Every Fibonacci sequence is associated to an infinite staircase, as explained in Section 4.1. The staircase has an average slope of $1/\tau$ (Figure 4.1(c)).

We now return to the problem of constructing candidates for one-dimensional crystals. First we will describe a construction method that *always* produces crystals. Then we will describe a method that produces them under certain well-understood conditions. Finally we will describe a method that produces crystals in some circumstances, about which a great deal remains to be understood.

4.3 Projected crystals

The 'projection method', introduced in Chapter 2, is inspired by Bohr's theorem that any almost periodic function is a section of a periodic (or limit periodic) function in some higher-dimensional space. It makes no use of any substitution property, and allows us to construct Delone crystals which do not have it. We will derive the Fibonacci crystals by projection and then discuss the method in greater generality.

We begin with the integer point lattice $\mathbb{Z}_2^p$; recall that its Voronoï cell is a square. Let ℓ be a line through the origin with slope $1/\tau$; it is a totally irrational subspace of E^2 (Figure 4.2), which means that it contains no other lattice points, nor does it pass through a vertex of the Voronoï cell of any lattice point.

Let X be the subset of $\mathbb{Z}_2^p$ whose cells are cut by ℓ. The points of X are unambiguously ordered by ℓ as we pass from one cell to the next: counting from the origin, the mth point of X is the vector $\vec{p}_m = (m_1, m_2)$, where $m_1, m_2 \in Z$, $m_1 + m_2 = m$, and the $(m+1)$st point is either $\vec{p}_{m+1} = (m_1 + 1, m_2)$ or $\vec{p}_{m+1} = (m_1, m_2 + 1)$ (Figure 4.3). Thus the points of X are the nodes of a unique staircase in $\mathbb{Z}_2^p$.

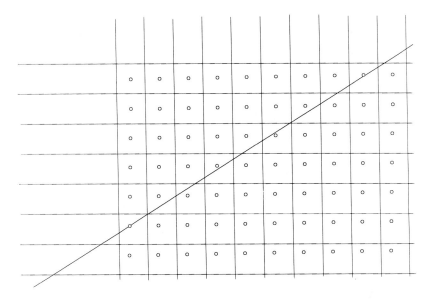

Fig. 4.2 The line is a totally irrational subspace of $\mathcal{I}_2$; X is the set of lattice points whose Voronoï cells are cut by the line.

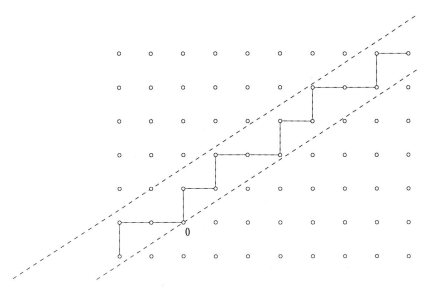

Fig. 4.3 The mth point of X is the vector $\vec{p}_m$ and the $(m+1)$st is $\vec{p}_{m+1} = \vec{p}_m + \vec{e}_j$, where $j = 1$ or $j = 2$.

Since the sequence $(\vec{p}_m)$ keeps track of the numbers of horizontal and vertical steps of the staircase, we can think of $\vec{p}_m$ as a population vector.

Letting Π be the orthogonal projector onto ℓ, we see that $\Pi(X)$ is a Delone set Λ with steps of lengths $|\Pi(\vec{e}_1)| = \tau/\nu$ and $|\Pi(\vec{e}_2)| = 1/\nu$ where, as before, $\nu^2 = \tau^2 + 1$.

In fact, we can find the coordinates of the points of Λ on ℓ explicitly. The line ℓ cuts the Voronoï cell of a lattice point (u, v) if and only if it intersects the principal (northwest to southeast) diagonal of the cell. The diagonals lie on the lines $x + y = m$, $m \in Z$; their intersections with ℓ, whose equation is $y = x/\tau$, are the points $(m/\tau, m/\tau^2)$. Thus ℓ cuts the cell of (u, v), where $u + v = m$, if and only if

$$u - \frac{1}{2} < \frac{m}{\tau} < u + \frac{1}{2} \quad \text{and} \quad v - \frac{1}{2} < \frac{m}{\tau^2} < v + \frac{1}{2}. \qquad (4.25)$$

Since $v = m - u$, the second inequality reduces to the first and (4.25) becomes

$$\left\| \frac{m}{\tau} \right\| = u,$$

where $\|x\|$ is the *nearest integer function*, assigning to the real number x the integer nearest to it (Appendix I). Thus the coordinates of the mth node of the staircase are

$$\vec{p}_m = (\|m/\tau\|, m - \|m/\tau\|). \qquad (4.26)$$

After projecting onto ℓ, we find that the mth point of $\Lambda = \Pi(X)$ has coordinates

$$x_m = \frac{m}{\nu} + \frac{1}{\tau\nu} \left\| \frac{m}{\tau} \right\|. \qquad (4.27)$$

By Proposition 2.18, the points whose Voronoï cells are cut by ℓ comprise the set

$$X = \{x \in \mathcal{I}_2^p | \Pi^\perp(x) \in K\}, \qquad (4.28)$$

where $\Pi^\perp$ is the orthgonal projector onto $\ell^\perp$, the line through the origin orthogonal to ℓ; $K = \Pi^\perp(V(0))$ is the window of the projection. It is easy to check that, in the Fibonacci case, K is the interval $[-\tau^2/2\nu, \tau^2/2\nu]$ on $\ell^\perp$.

Of course, we still have to prove that the projected sequence is a Fibonacci sequence.

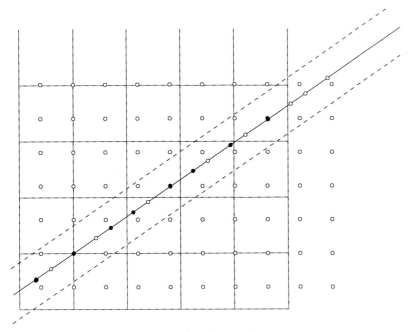

Fig. 4.4 A predecessor (solid circles) for the projected set a (solid and open circles on *l*) is projected from a magnified point lattice.

Proposition 4.4 $\Lambda = \Pi(X)$ is a Fibonacci sequence.

Proof We have to show that the difference sequence $\Pi(\vec{p}_{m+1} - \vec{p}_m)$ has predecessors at all levels. This means that we must show that there is a rescaled copy of $\mathcal{I}_2^p$ that plays the same role with respect to the predecessor of Λ that $\mathcal{I}_2^p$ plays with respect to Λ; then the same argument can be used to construct *its* predecessor, and so on *ad infinitum*. The lattice generated by the vectors $\tau\vec{e}_1$ and $\tau\vec{e}_2$ is $\tau\mathcal{I}_2$ and it is easy to check that its Voronoï cell projects to τK on $\ell^\perp$ (Figure 4.4). We must also show that the new construction does not introduce any new points, that is, that all the points projected from $\tau\mathcal{I}_2^p$ are also projected from $\mathcal{I}_2^p$. A point of $\tau\mathcal{I}_2^p$ has coordinates $(\tau u, \tau v)$, where $(u, v) \in \mathcal{I}_2^p$; its projection onto ℓ is $(\tau^2 u + \tau v)/\nu$. We will show that this is also a point of $\Pi(X)$. Since

$$\tau^2 u + \tau v = \tau(u + v) + u = (u + v, u) \cdot (\tau, 1),$$

we have to show that $\nu\Pi^\perp(u + v, u) \in [-\tau^2/2, \tau^2/2]$. This follows easily from the fact that $\nu\Pi^{\perp}(u, v)$ lies in this interval and

$$\nu\Pi^{\perp}(u + v, u) = -(u + v) + \tau u = \frac{1}{\tau}(u - \tau v).$$

❏

The window of a canonical projection contains a great deal of information about the projected sequence. For example, we can use it to determine the frequencies of elements of the (τ/ν) – atlas of the Fibonacci sequence (recall Definitions 1.6 and 1.8). The same type of argument will be used to compute the frequencies of the elements of atlases in projected tilings (Chapters 5 and 6).

The Fibonacci (τ/ν) – atlas has three elements, corresponding to the three admissible words of length two in a Fibonacci string. Let us say that a point x_k is of adjacency type AA if it lies between two segments of length τ/ν and of type AB if it lies between segments of lengths τ/ν (on the left) and $1/\nu$ (on the right); the adjacency type BA is defined similarly.

To compute the frequencies of the elements of the atlas, we note that the four edges of $V(0)$ project to four subintervals of K, and these subintervals partition K into three subintervals with disjoint interiors (Figure 4.5). Omitting the factor $1/\nu$ for the moment, the subintervals are:

$$U_1 = [-\tau^2/2, (1 - \tau)/2],$$

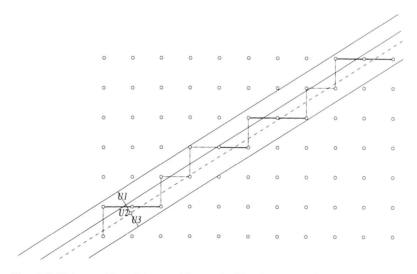

Fig. 4.5 K is partitioned into subintervals U_1, U_2, U_3; each corresponds to a different adjacency type (BA, AA, or AB) in the tiling.

$$U_2 = [(1 - \tau)/2, (\tau - 1)/2],$$

$$U_3 = [(\tau - 1)/2, \tau^2/2].$$

We have $|U_1| = |U_3| = 1$ and $|U_2| = \tau - 1$. Thus $|U_2 \cup U_3| : |U_1| = \tau : 1$.

Proposition 4.5 The point x_k is of adjacency type AA if and only if $\Pi^\perp(x_k) \in U_2$ (Figure 4.5).

Proof A lattice point $(u, v) \in X$ projects to a point of type AA if and only if $\Pi^\perp(u - 1, v)$, $\Pi^\perp(u, v)$, and $\Pi^\perp(u + 1, v)$ all lie in the interval $K = [-\tau^2/2, \tau^2/2]$. Thus u and v must satisfy the three inequalities

$$-\tau^2/2 \le -u + \tau v \le \tau^2/2,$$

$$-\tau^2/2 - 1 \le -u + \tau v \le \tau^2/2 - 1$$

and

$$-\tau^2/2 + 1 \le -u + \tau v \le \tau^2/2 + 1.$$

It follows that

$$-u + \tau v \in [(1 - \tau)/2, (\tau - 1)/2] = U_2.$$

Since the length of U_2 is $\tau - 1 = 1/\tau$ and $\Pi^\perp(X)$ is uniformly distributed in K (see Chapter 2), the frequency of this configuration is

$$|U_2|/|K| = (1/\tau)/\tau^2 = \tau^{-3}.$$

Similarly, we can identify the adjacency types AB and BA with the intervals U_3 and U_1 respectively.

In a similar way, we can show

Proposition 4.6 The relative number of long and short intervals in $\Pi(X)$ is $\tau : 1$.

Proof Let $y_{n+1} = x_{n+1} - x_n$ where $x_n = \vec{\Pi}(p_n)$. We will show that $y_{n+1} = \tau$ if and only if $(p_n) \in U_2 \cup U_3$; otherwise $y_{n+1} = 1$. Now $\Pi(p_{n+1} - p_n) = \tau$ if and only if ℓ cuts the Voronoï cells of both (u, v) and $(u + 1, v)$. Both of these points must project into K, and it is easy to see that this means

$$-u + \tau v \in U_2 \cup U_3.$$

$\square$

This proposition also follows from Proposition 4.1.

4.3.1 Projected Fibonacci sequences are crystals

We will outline three very different proofs of the fact that Fibonacci sequences are crystals; all are based on the projection method. In Section 4.4 we will draw the same conclusion from their substitution properties.

First proof. We will show that a projected Fibonacci sequence is a periodically modulated point lattice, and then apply Theorem 3.3. We begin with (4.27) (and the properties of $||x||$ and the greatest integer function $[x]$ discussed in Appendix I). When x is irrational, we have

$$||x|| = [x + \frac{1}{2}].$$

Since $[x] = x - \{x\}$, where $\{x\}$ is the 'fractional' part of x, we can write (4.27) in the form

$$x_m = \frac{m}{\nu} + \frac{1}{\nu\tau}\left(\frac{m}{\tau} + \frac{1}{2} - \{\frac{m}{\tau} + \frac{1}{2}\}\right)$$
$$= \frac{m\nu}{\tau^2} + \frac{1}{\nu\tau}\left(\frac{1}{2} - \{\frac{m}{\tau} + \frac{1}{2}\}\right).$$

The terms $m\nu/\tau^2$ are the points of a linear lattice; in fact they are precisely the points at which the parallel, equally spaced lines $x + y = m$ intersect ℓ. The terms

$$\frac{1}{\nu\tau}\left(\frac{1}{2} - \{\frac{m}{\tau} + \frac{1}{2}\}\right)$$

are values, for integer arguments, of the periodic function

$$\frac{1}{\nu\tau}\left(\frac{1}{2} - \{\frac{x}{\tau} + \frac{1}{2}\}\right),$$

which has period τ and varies from $-1/2\nu\tau$ to $1/2\nu\tau$. Thus (Figure 4.6) the projected points are located 'near' the points of the one-dimensional point lattice

$$m\nu/\tau^2, \quad m \in Z.$$

The hypotheses of Theorem 3.3 are satisfied and so

Proposition 4.7 The Fibonacci sequence is a crystal.

Three approximations to the graph of the intensity function of this crystal are shown in Figure 3.10; a computer simulated diffraction pattern is shown Figure 4.7. (For a 'real' Fibonacci diffraction pattern, see (Litvin, Romberger, and Litvin, 1988).)

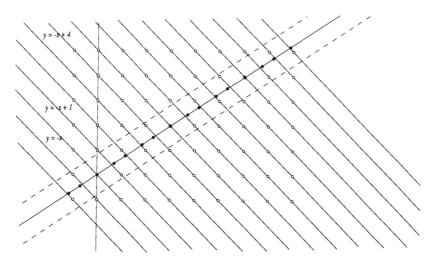

Fig. 4.6 The projected points (solid circles) oscillate about a linear point lattice (the points where the lines $y = -x + n$ intersect ℓ).

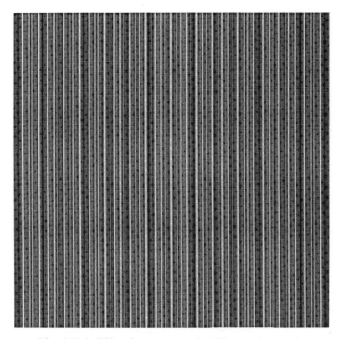

Fig. 4.7 A diffraction pattern of a Fibonacci crystal.

Second proof. This proof makes use of the point lattice dual to the one in which the staircase is a path. We want to study the sum

$$\hat{\rho}(s) = \sum_{x_m \in F} \exp(-2\pi i x_m s). \tag{4.29}$$

The population vector can be written either in the form

$$\vec{p}_m = m_1 \vec{e}_1 + (m - m_1) \vec{e}_2$$

or in the form

$$\vec{p}_m = x_m \vec{\epsilon}_1 + y_m \vec{\epsilon}_2 \tag{4.30}$$

where $\vec{\epsilon}_1$ is a unit vector on ℓ, $\vec{\epsilon}_2$ is a unit vector orthogonal to $\vec{\epsilon}_1$, and y_m is the projection of $\vec{p}_m$ on $\ell^\perp$: y_m lies in the window K
   Setting

$$\vec{w} = s\vec{\epsilon}_1 + t\vec{\epsilon}_2 \tag{4.31}$$

we have

$$\vec{p}_m \cdot \vec{w} = x_m s + y_m t. \tag{4.32}$$

Then (4.29) becomes

$$\hat{\rho}(s) = \sum_{m \in Z} \exp(-2\pi i \vec{p}_m \cdot \vec{w}) \exp(2\pi i y_m t). \tag{4.33}$$

If $\vec{w}$ is a vector in the dual lattice (which is again $\mathcal{I}_2$, since the staircase is drawn in $\mathcal{I}_2^p$ and $\mathcal{I}_2$ is self-dual), then $\vec{p}_m \cdot \vec{w} \in Z$ and $\exp(-2\pi i \vec{p}_m \cdot \vec{w}) = 1$. Thus

$$\hat{\rho}(s) = \sum_{y_m \in K} \exp(2\pi i y_m t); \tag{4.34}$$

equation (4.32) shows how t depends on s. So when $\vec{w}$ is in the dual lattice, we can work with (4.34).
   By (3.45), we want to show that

$$\lim_{N \to \infty} \frac{1}{N} \sum_{m=-N}^{N} \exp(2\pi i y_m t)$$

is a nonzero constant; this follows from the uniform distribution of the projection of $\mathcal{I}_2^p$ onto $\ell^\perp$. Thus $\hat{\rho}(s)$ has a delta at every s satisfying (4.32) with $\vec{w} \in \mathcal{I}_2$. (See Elser, 1986; de Bruijn and Senechal, 1994.)

To see which points these *s*s actually are, we write $\vec{w}$ in terms of the $\vec{e}_1, \vec{e}_2$ basis:

$$\vec{w} = (u, v), \quad u, v \in Z.$$

Then, by (4.31),

$$s = \frac{1}{\nu}(u, v) \cdot (\tau, 1) = \frac{u\tau + v}{\nu}.$$

The numbers $u\tau + v$ are the integers of the algebraic number field $Z[\tau]$ (Appendix I). This set of points is an orbit of a Z-module of rank 2; it is dense in the real line.

Third proof. The second proof gives no information about whether there is also a continuous component in the spectrum, but the third one does. Let ρ_X, ρ_L be the densities of X (Eq. (4.28)) and I_2^p and let χ_C be the characteristic function of the cylinder C. Then

$$\rho_X = \rho_L \chi_C$$

and

$$\hat{\rho}_X = \hat{\rho}_L * \hat{\chi}_C.$$

By the Poisson summation formula,

$$\hat{\rho}_L = \rho_L$$

and one can show that this convolution is a Dirac comb. (For details, see Katz and Duneau (1986); for a proof that their computations can be justified for the generalized functions involved, see Porter, (1988).) De Bruijn (1986b) reaches the same conclusion by showing that $\Pi(X)$ is a Poisson comb.

4.3.2 Generalizations

We can generalize the projection method to produce uncountably many different Fibonacci sequences, and also uncountably many other aperiodic crystals.

To obtain the other Fibonacci sequences, we use translates $\ell + \vec{t}$ of ℓ, $\vec{t} \in E^2$, to cut the Voronoï cells of I_2^p; this is equivalent to translating the cylinder $C = \mathcal{E} \oplus \ell$ along $\ell^\perp$. The construction is unambiguous except in *singular* cases, when $\ell + \vec{t}$ passes through a vertex of a Voronoï cell; the singular sequences can be treated as limiting cases of *regular* ones. All of the properties of the projected set discussed in the preceding subsections hold when ℓ is replaced by $\ell + \vec{t}$, if $\vec{t}$ is regular.

Indeed, all of these properties – except those related to substitution – hold when ℓ is replaced by *any* line with irrational slope and $\mathcal{I}_2$ is replaced by $\mathcal{I}_n$.

We conclude that

Proposition 4.8 One-dimensional projected Delone sets are crystals.

4.4 Substitution sequences

Since we have already studied the substitution properties of Fibonacci numbers, we will only indicate how they too imply that Fibonacci sequences (and many other substitution sequences) satisfy the diffraction condition. This section is based largely on the work of Bombieri and Taylor (1987).

Let us return to the sequence of finite strings (4.22). By identifying the As with intervals of length τ and the Bs with intervals of length 1, we obtain a sequence of finite colinear point sets; the coordinate of the right-most point of the kth finite sequence is x_{F_k}. These point sets can, in turn, be represented by densities which are finite sums of deltas. Denoting these densities by $\rho_0(x), \ldots, \rho_k(x), \ldots$ and the density of their limit, the infinite Fibonacci sequence, by $\rho(x)$, we have

$$\lim_{k \to \infty} \rho_k(x) = \rho(x)$$

and

$$\lim_{k \to \infty} \hat{\rho}_k(s) = \lim_{k \to \infty} \sum_{j=0}^{F_k} \exp(-2\pi i x_j s) = \hat{\rho}(s).$$

This limit is a case of (3.45), where the subsets Λ_i are the 'Fibonacci subsequences' of the sequence of partial sums. By (4.14),

$$x_{F_{k+1}} = (F_k, F_{k-1}) \cdot (\tau, 1);$$

by (4.20), we know that the lattice points (F_k, F_{k-1}) eventually lie in a cylinder parallel to the line ℓ. Thus we are back to the staircase and the Fibonacci sequence is a crystal.

As closer look at this argument shows, however, that it depends on the fact that one eigenvalue of the substitution matrix is expansive and the other contractive. If this were not the case – and generically it is not – we could not conclude that the sequence is a crystal.

To get a picture of 'what is going on', we consider more general substitution rules. An $n \times n$ matrix $\mathcal{A}$ is said to be *primitive* if its entries are nonnegative integers and if there exists a positive integer k such that all the entries of $\mathcal{A}^k$ are positive. We can think of $\mathcal{A}$ as the matrix of a rule that tells us how to replace n symbols $\{A_1, \ldots, A_n\}$ by finite strings of these same symbols (Section 5.3); the requirement of primitivity ensures that every symbol will appear after $k + 1$ generations, no matter what initial sequence we choose.

Definition 4.3 A substitution rule is a linear map that can be represented by a primitive matrix.

Every substitution rule defines a family of infinite substitution strings that can be associated to one-dimensional Delone sets.

Primitive matrices have another important property: a primitive matrix has a leading eigenvalue, one that is strictly greater, in absolute value, than all the others (this is the Perron–Frobenius theorem – see Appendix I). This property guarantees that the symbols in the infinite string appear with well-defined frequencies, because it means that asymptotically the population vector tends to the direction of the (left) eigenvectors belonging to the leading eigenvalue. (The important distinction between left and right eigenvectors is also discussed in Appendix I.) The Perron–Frobenius theorem does *not* imply that the smaller eigenvalues are contractive, that is, that they are less than one in absolute value.

For an example of how sequences without the contractive property can behave, we consider a substitution sequence that looks, at first sight, very much like the Fibonacci sequence, except that both eigenvalues are expansive (Lançon and Billard, 1992). The substitution rule is

$$A \to AABA, \quad B \to BAB, \tag{4.35}$$

with matrix and characteristic equation

$$Q = \begin{pmatrix} 3 & 1 \\ 1 & 2 \end{pmatrix}, \quad \lambda^2 - 5\lambda + 5 = 0. \tag{4.36}$$

The eigenvalues of Q are $\lambda_1 = 2 + \tau \sim 3.618$ and $\lambda_2 = 3 - \tau \sim 1.382$, but the eigenvectors of Q are exactly the same as in the Fibonacci case and so the ratio of As to Bs in the infinite strings is also $\tau : 1$.

Thus these strings – let us call them LB strings – can also be associated with Delone sets with step sizes proportional to τ and 1, and with staircases whose average slope is $1/\tau$. But in this case the population vector of the mth generation,

$$\vec{p}_m = \frac{(\tau + 2)^m}{\nu} \epsilon_1 + \frac{(3 - \tau)^m}{\nu} \epsilon_2,$$

expands in both directions.

An *LB* staircase is shown in Figure 4.2(b).

Godrèche and Luck (1992) have shown that the maximal intensities associated to *LB* sequences are not comparable to Dirac deltas, but are in the range expected for singular continuous spectra (see also Godrèche, 1990).

Returning to general substitution sequences, we note that when $\mathcal{A}$ is an $n \times n$ matrix, the 'population vector' $\vec{p}_m$ will be a vector in $\mathcal{I}_n$:

$$\vec{p}_m = \vec{p}_0 \mathcal{A}^n. \tag{4.37}$$

By the Perron–Frobenius theorem, the relative frequencies of the symbols in successive generations approach limiting values in the infinite string. We associate line segments to these letters, choosing their lengths $\alpha_1, \ldots, \alpha_n$ so that their ratios are preserved under substitution; this will be the case when $\vec{\epsilon}_1 = (\alpha_1, \ldots, \alpha_n)$ is a right eigenvector belonging to the leading eigenvalue λ_1 of $\mathcal{A}$. Finally, we define an $\mathcal{A}$ sequence to be any sequence of points $\Lambda = \{x_n\}$ such that, for all n, $x_n - x_{n-1} \in \{\alpha_1, \ldots, \alpha_n\}$ and Λ has suitably defined predecessors of all orders with respect to the substitution $\mathcal{A}$.

As we have already seen, whether the diffraction condition is satisfied depends on the subtle properties of the eigenvalues of $\mathcal{A}$. We now describe these properties in more detail.

Let μ_1 be a real algebraic integer of degree k, and $\mu_2, \ldots, \mu_k$ its algebraic conjugates (Appendix I.) The leading eigenvalue of a primitive matrix is always a real algebraic integer.

Definition 4.4 An algebraic integer μ_1 of degree k is said to be a Pisot – Vijayvaraghavan (or PV) number if $\mu_1 > 1$ and $|\mu_j| < 1$, $j = 2, \ldots, k$.

For example, τ is a PV number, since its conjugate is $-1/\tau$, but $2 + \tau$ is not, since its conjugate is $3 - \tau$.

The importance of PV numbers for the diffraction condition lies in an alternative characterization of PV numbers (Pisot, 1946; Cassells, 1965).

Theorem 4.1 Let $\mu_1 > 1$ be a real algebraic integer. Then μ_1 is a PV number if and only if there exist nonzero $q \in R$ such that

$$\lim_{m \to \infty} \mu_1^m q = 0 \pmod{Z}. \tag{4.38}$$

It follows that

Theorem 4.2 The diffraction condition is satisfied for an $\mathcal{A}$ sequence if and only if the leading eigenvalue λ_1 of $\mathcal{A}$ is a PV number.

Proof (sketch) First let us assume that the leading eigenvalue λ_1 is PV. Then for sufficiently large m the population vector of the mth generation is

$$\vec{p}_m \sim \lambda_1^m \vec{p}_0$$

and we can write

$$\hat{\rho}_m(s) \sim 1 + \cdots + \exp(-2\pi i \lambda_1^m \vec{p}_0 \cdot \vec{\epsilon}_1 s). \tag{4.39}$$

By (4.38) we have s such that

$$\lim_{m \to \infty} \frac{1}{m} \hat{\rho}_m(s) = c(s) \neq 0 \tag{4.40}$$

(These s lie in a certain countable subset of the algebraic field of λ_1.) Thus the diffraction condition is satisfied.

Conversely, if the diffraction condition is satisfied, (4.40) must hold for a countable infinity of values of s. As Bombieri and Taylor pointed out, the characteristic equation of $\mathcal{A}$ has an associated recurrence relation, and the sequence $(\vec{p}_n \cdot \vec{\epsilon}_1)$ corresponding to the successive generations of symbols also obeys this recurrence. The recurrence carries over to $\rho_n(s)$ and $\hat{\rho}_n(s)$.

We will show how the computation works for the Fibonacci case; the general case is more complicated but the principle is the same. The recurrence relation for the Fibonacci sequence is (4.8). It follows that

$$\sum_{j=0}^{F_{m+1}} \delta(x - x_j) = \sum_{j=0}^{F_m} \delta(x - x_j) + \sum_{j=0}^{F_{m-1}} \delta(x - x_{F_m} - x_j).$$

This implies that

$$\rho_{m+1}(x) = \rho_m(x) + \rho_{m-1}(x - x_{F_m}), \tag{4.41}$$

where $\rho_m(s)$ is the density associated to the mth generation. Since Fourier transformation is linear, we have

$$\hat{\rho}_{m+1}(s) = \hat{\rho}_m(s) + \sum_{j=0}^{F_{m-1}} \exp(-2\pi i (x_{F_m} + x_j)s)$$

$$= \hat{\rho}_m(s) + \exp(-2\pi i (x_{F_m}s))\hat{\rho}_{m-1}(s). \tag{4.42}$$

We divide both sides of (4.42) by m and study the limit as $m \to \infty$.

Equations (4.41) and (4.42) are consistent if and only if

$$\lim_{m\to\infty} \exp(-2\pi i x F_m s) = \lim_{m\to\infty} \exp(-2\pi i \lambda^m \vec{p}_0 \cdot \vec{\epsilon}_1 s) = 1;$$

that is, if λ_1 is a PV number.

4.5 Circle map sequences

There is at least one more construction for the Fibonacci crystal, the *circle map* method, introduced into quasicrystallography by Aubry and Godr̆eche (1986). It is especially interesting because it generates Delone sets whose diffraction properties are not completely understood.

We start with a circle C of circumference (not radius) one, and divide it into two arcs,

$$D = [0, 1/\tau^2), \qquad 1 - D = [1/\tau^2, 1);$$

0 and 1 are of course identified. Notice that $|1 - D|/|D| = \tau$. Now take a line marked with the points of a linear point lattice of spacing $1/\tau^2$; choosing a lattice point as origin, the coordinate of the nth point is

$$q_n = n/\tau^2.$$

We place q_0 at the point 0 on C and wind the line around C counterclockwise (Figure 4.8).

This defines a map $q_n \to c_n$ from the point lattice to a set of points on the circle; note that

$$c_n \in \begin{cases} D, & \text{if } 0 \leq \{n/\tau^2\} < 1/\tau^2, \\ 1 - D, & \text{if } 1/\tau^2 \leq \{n/\tau^2\} < 1; \end{cases} \qquad (4.43)$$

as usual, $\{x\}$ denotes the fractional part of x.

We can replace the sequence (c_n) by a sequence of As and Bs, according as $c_n \in 1 - D$ or $c_n \in D$. Since the sequence (q_n) is uniformly distributed modulo one (see Appendix I), the ratio of As to Bs will be approximately $(1 - D)/D = \tau$.

We can also interpret the sequence q_n as a staircase in the integer point lattice $\mathcal{I}_2^p$ by the map

$$q_n \to \begin{cases} \vec{e}_1, & \text{if } c_n \in 1 - D, \\ \vec{e}_2 & \text{otherwise.} \end{cases}$$

This amounts to changing the scale of the linear lattice and bending it at right angles as dictated by the circle map. Let H be the set of

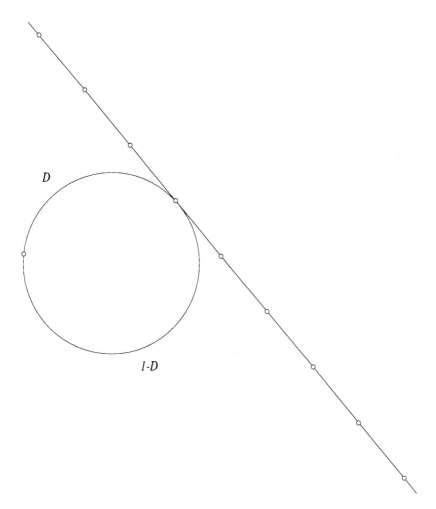

Fig. 4.8 The circle C is partitioned into arcs D and $1 - D$, and a linear lattice is mapped onto the circle.

staircase nodes and ℓ the line through the origin with slope $1/\tau$. Projecting H onto ℓ, we obtain a sequence of points $\{x_n\}$ with two spacings, τ and 1, the lengths of the projections of $\vec{e}_1$ and $\vec{e}_2$ onto ℓ. Since the length of D is $1/\tau^2$, the spacings of length 1 are never successive, and since $2/\tau^2 > 1 - 1/\tau^2 > 1/\tau^2$, there can never be more than two successive spacings of length τ.

Proposition 4.9 The sequence (x_n) is a Fibonacci sequence.

The proof is left as an exercise.

The circle map construction is very general. D can be any real number between 0 and 1 and the spacing of points in the linear point lattice can be any irrational number γ. We can also place the origin of the linear lattice at any point on the circle. We can always construct a staircase H in $\mathcal{I}_2^p$ and project it onto the line through the origin with slope $(1 - D)/D$, as above. But does the projected sequence always satisfy the diffraction condition?

The answer is subtle: the construction does not guarantee that $\Pi^\perp(H)$ is contained in any interval in $\ell^\perp$. It does guarantee (because the sequence (c_n) is uniformly distributed on the circle) that the staircase has an average slope and thus asymptotically the two steps appear with well-defined frequencies. But just as in the case of substitution sequences, we have to look more closely at the discrepancy or fluctuation of the component of the population vector $\vec{p}_m$ in to determine whether the projected point set is a crystal.

Let us write $\vec{p}_m = (m_D, m - m_D)$, where m_D is the number of lattice points among the first m that are mapped into D. After rescaling by the length of the vector $(D - 1, D)$, the length of the projection of $\vec{p}_m$ onto $l^\perp$ is

$$|(m_D, m - m_D) \cdot (D - 1, D)| = |m_D(D - 1) + (m - m_D)D| = |-m_D + mD|.$$

Under what conditions do these values lie in a bounded set? The answer is given by a theorem of Kesten (1966):

Theorem 4.3 (Kesten) There is a $c > 0$ such that $|-m_D + mD| < c$ when m is sufficiently large, if and only if $D = j\gamma$, where $j \in Z$.

In the special case of the Fibonacci sequence, $D = \gamma = 1/\tau^2$, so $j = 1$; in fact, the canonical projection from $\mathcal{I}_2^p$ to an irrational line is *always* a circle map sequence with $j = 1$ (see Senechal, 1994).

It is not known whether the Kesten condition is necessary or sufficient for the diffraction condition to hold for $\Pi(H)$. It would be a sufficient condition if it implied the uniform distribution of the projected staircase nodes in the window.

Godrèche, Luck, and their colleagues (Aubry, Godrèche, and Luck, 1988; Godrèche, Luck and Vallet, 1987) have studied the spectra of circle map sequences that do not obey the Kesten condition (for example, when $D = 1/2$). They concluded, on the basis of numerical methods, that the diffraction spectra of these Delone sets are singular continuous.

4.6 Staircases

The general question, *what are the defining characteristics of nonperiodic one-dimensional Delone sets that satisfy the diffraction condition?*, is still far from being answered. Only in the case of canonical projection is the relation between local order and diffraction fairly clear. In this case, because the lattice points that lie in the cylinder are uniformly distributed in the window, we know that

(i) every local pattern must appear with a frequency close to its limiting one in every sufficiently large interval in ℓ, and
(ii) the projected set is a crystal.

Is (i) necessary or sufficient for (ii)? In any case, property (i) does not hold for staircases in general, even those that lie in some cylinder: when the projection is not canonical, the nodes of a staircase may be a subset of the lattice points in the cylinder in which they lie. If their projections onto $\ell^\perp$ are uniformly distributed in the window, then the projected points will be a crystal by the second argument in Section 4.3.1, but it may no longer be possible to identify each subinterval of the window with a single local pattern.

For still more general staircases there is very little we can say. Even if we restrict ourselves to the case where the number of step sizes is two, there are far more questions than answers. After all, any infinite string of two letters A and B can be associated with a staircase, and the nodes of the staircase can be projected to any line. But we can pose some interesting questions that are more specific. For example:

• What happens to the spectrum of a substitution sequence if we modify the local geometry of the substitution?

• What can we say about the spectra of Delone sequences generated in other interesting ways, such as paperfolding or wire bending?

• To what extent is the diffraction condition robust? For example, does this property continue to hold when the slope of the line onto which the staircase is projected is changed?

• How 'amorphous' can a staircase sequence be? We have seen that there are sequences with purely discrete spectra, and evidently there are sequences with purely singular continuous spectra. Do there exist staircase sequences with purely absolutely continuous spectra?

To connect these problems to the fluctuations of the staircase in $\ell^\perp$, we again let H be the set of staircase nodes, and study the closure of $\Pi^\perp(H)$ in K. This set is sometimes called the *atomic surface* associated with the

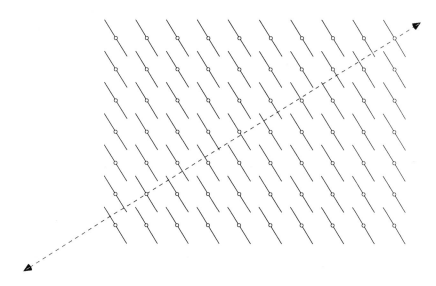

Fig. 4.9 In the canonical projection the atomic surface is a line segment.

staircase; the original sequence can be retrieved as the intersection of ℓ with copies of the surface placed at the points of $\mathcal{I}_2^p$.

In the case of canonical projection from E^2 to ℓ the atomic surface is a line segment (Figure 4.9). In more general situations the structure of the atomic surface can be quite complicated. For example, Luck, Godrèche, Janner, and Janssen (1993) have studied how the surface is affected by interchanging the letters in a substitution rule (which, after all, only governs the number of letters of each kind in the substitution, not the order in which they occur). In some cases, the surfaces appear to have fractal boundaries.

Paperfolding sequences are discussed in Mendès France and van der Poorten (1981) and Dekking, Mendès France, and van der Poorten (1982a; 1982b); for wire-bending sequences, see Mendès France and Shallit (1989).

4.7 Notes

1. Our staircases are reminiscent of Coxeter's 'orthogonal trees' (Coxeter, 1969); our computations of local patterns are analogous to the 'safemoves' in 'Wythoff's game' (Ball and Coxeter, 1974; Porta and Stolarsky, 1990).

2. A few miscellaneous remarks on the construction methods:

Fibonacci crystals are also discussed in Senechal (1991a). A discussion of the technical sense in which a Fibonacci string is a limit of finite strings is beyond the scope of this book; see e.g. Queffelec (1987).

The projection technique described here is based on de Bruijn (1981a); it has become the standard method for constructing nonperiodic point sets and tilings in every dimension.

One very important question that we have not addressed explicitly is: how are the various construction methods related to one another? Specifically, which projected sequences have the substitution property, and vice versa? It is clear that canonical projection does not imply substitution, since if the slope of ℓ is a transcendental number instead of an algebraic one, then the projected sequence cannot be a substitution sequence. On the other hand, Bombieri and Taylor have shown that every substitution sequence with the Pisot property is a subset of a canonically projected sequence, but it is not necessarily one itself. The very interesting paper by Lunnon and Pleasants (1987) addresses some of these questions; it deserves more attention than it has received (here or elsewhere).

The circle map discussed in this section was also used in the early stages of developing a model of cardiac arrythmia (Glass, Goldberger, and Belair, 1986).

3. It is a curious and interesting fact that although regular (canonically) projected sequences are Poisson combs, there is a continuous component in the spectrum in singular cases. This is because, if a vertex of a Voronoï cell lies on the boundary of the cylinder, so does its image through the center of the cell and thus we introduce a new step size (the body diagonal of the cell). The density $\rho(x)$ will no longer be a Poisson comb because the (single) term contributed by this step contributes a continuous component to $\hat{\rho}(s)$ (see e.g. de Bruijn (1986b) and Hof (1993).)

4. The diffraction spectrum is not the only way to measure long-range order in a one-dimensional Delone set. For some other interesting approaches to this question, see Mendès France (1984) and Allouche and Mendès France (1986). Also, it is sometimes advantageous to use the broader definition of spectrum that is generally employed for symbolic dynamical systems.

5. The work of Luck, Godrèche, Janner, and Janssen shows us some of the subtleties that we can encounter in working with even simple, deterministic sequences. As another example, consider the famous Thue-Morse sequence, generated by the substitution

$$0 \longrightarrow 01, \quad 1 \longrightarrow 10. \tag{4.44}$$

If we begin with a single digit (0th generation), the nth generation will have 2^n digits. For example, beginning with a single 0, the fifth generation is

$$01101001100101101001011001101001$$

In every generation, the ratio of 0s to 1s is 1:1; nonperiodicity follows from Theorem 5.3.

Mahler (1926) showed that the spectrum – of the translation operator – on the space of Thue-Morse sequences is singular continuous. However, van Enter and Miękisz (1992) have shown that that spectrum becomes discrete when its elements are grouped together in a certain way. Notice also that if we interpret the sequence as a pattern of vacancies in a one-dimensional point lattice (1 = atom present, 0 = vacancy) then the diffraction pattern necessarily has a discrete component, as we showed in Chapter 3.

6. There is an enormous literature on the geometry of sequences which deserves a review of it own. See, for example, Allouche (1987), Lunnon and Pleasants (1992), Porta and Stolarksy (1990), Series (1985), Mendes France (1988), and Queffelec (1987); Queffelec's book contains a large bibliography.

7. The role of uniform distribution in the computation of diffraction patterns was first pointed out by Elser (1986).

5

Tiles and tilings

It seems a mystery to me that we do need a whole space to piece
the neighborhoods together.

(S. S. Chern)

5.1 Tilings and crystals

Definition 5.1 (informal) A tiling is a partition of a space into a countable
number of tiles (Figure 5.1).

Tiling theory is an ancient subject, with roots in natural philosophy and
the practical arts. It is concerned with the ways in which a surface or space
can be filled by copies of shapes of a few kinds, without overlapping or
leaving gaps, and seeks to answer questions about the properties of the
patterns created in this way. In science, tilings arise in several different

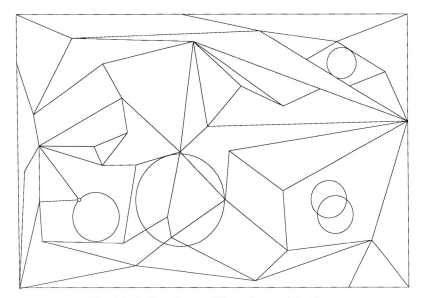

Fig. 5.1 A tiling is a partition of space into tiles.

contexts. For example in biology one studies the network of cells created by cell division, and in solid state science one studies how crystals grow as atomic units arrange themselves into some sort of pattern. In these and other cases one is interested in the rules that govern the processes, and the properties of the tilings that the rules produce.

In the last decade or so, tiling theory has emerged from papers scattered in the journals of many sciences, from 'recreational mathematics', and from the world of Escher enthusiasts, to assume an important place in contemporary mathematical research. This welcome development is due to several factors internal to mathematics in addition to its importance in the sciences. Tiling theory was given coherence by Grünbaum and Shephard (1987), who clarified and unified the theory of tilings of the plane and laid a theoretical foundation for much of it. New directions that link tilings with other parts of mathematics have been pioneered by Conway, Thurston, and others (see Thurston (1989) for an account of some of these developments). With computers we can draw tilings to order and then modify them interactively; this new ability to visualize and experiment is revolutionizing the subject. But one of the greatest stimuli for recent research has been the fascinating questions posed by quasicrystals.

In this chapter we review some of the basic concepts of tiling theory and some important methods for constructing tilings. Of course, we will emphasize those parts of the subject that deal with nonperiodic tilings and are (possibly) relevant to quasicrystals, including their diffraction patterns.

We will refer to Grünbaum and Shephard (1987) so often that we will abbreviate this reference by G&S; we will also use this abbreviation to refer to these two authors when we are discussing material that appears in that book.

5.2 Some basic concepts of tiling theory

We need a more formal definition of a tiling than Definition 5.1, one which states its defining features explicitly: every point of the space belongs to at least one tile, and if a point belongs to two or more, it lies on their common boundary.

Definition 5.2 A tiling $\mathcal{T}$ of the space E^n is a countable family of closed sets called tiles:

$$T = \{T_1, T_2, \ldots\}$$

such that int $T_i \cap$ int $T_j = \emptyset$ if $i \neq j$ and $\cup_{i=1}^{\infty} T_i = E^n$.
(As usual, int T means the interior of T.)

To avoid pathologies, we assume that the tiles have positive volume and are the closures of their interiors. So, for example, a circle cannot be a tile in E^2 but a disk can be, and while a line segment can be a tile in E^1, it cannot be a tile in a space of any higher dimension.

Definition 5.2 does not restrict the shapes of the tiles in any way, nor does it limit the number of different shapes. In some contexts, such as the study of cell networks in biological tissues, it would be inappropriate to place any such restrictions, since no two cells are likely to be congruent, but it may be useful to make some assumptions about the 'skeleton' of the tiling (its network of vertices and edges). In other contexts, such as crystallography, it is appropriate to assume that the tiles have only a few shapes – possibly only one – but one does not want to place any *a priori* restrictions on the skeleton.

Since this book is inspired by problems in crystallography, we will consider tilings whose tiles are copies of a finite set of 'prototiles'; in other words, the tiles belong to a finite set of equivalence classes. Exactly what equivalence means depends on the context. For example, a tiling of a floor by identical squares has just one prototile: equivalence means congruence. But to describe a checkerboard properly we should say there are two prototiles: both have square shapes but we distinguish them by colors. To construct the checkerboard, we also need a 'matching rule' for putting the tiles together (no square can share an edge with one of the same color); this rule must be part of the definition of 'prototile', too. As another example, the tiling of Figure 5.2 has one or two prototiles, depending on whether 'equivalence' means congruence, or only translation.

Definition 5.3 Let $\{T_1, T_2, \ldots\}$ be the tiles of a tiling T, partitioned into a set of equivalence classes by some criterion $\mathcal{M}$. A set P of representatives of these classes is called the protoset for T (with respect to $\mathcal{M}$).

$\mathcal{M}$ may be a transformation group or decorations of the tiles (see Section 5.5), or both. Normally, the matching rules can be realized as deformations of the edges of the tiles (see Chapter 7). For example, instead of coloring the squares of a chess board red and black, we might deform them as shown in Figure 5.3. But it is not always helpful to do this,

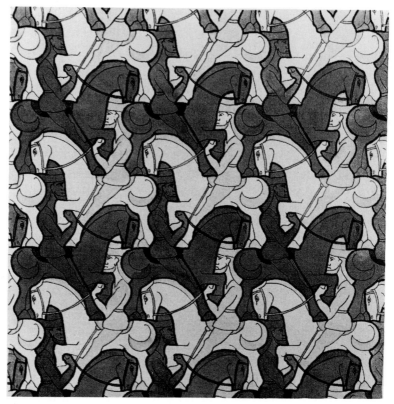

Fig. 5.2 Does this tiling have one prototile or two? (M. C. Escher Foundation – Baarn – Holland. All rights reserved.)

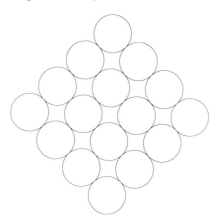

Fig. 5.3 A version of the chessboard in which colors are not needed: the circles represent the squares of one color and the interstices the squares of the other.

as important insights can be lost. In this chapter we will describe $\mathcal{M}$ explicitly, but remember that it is always somewhere in the background.

Definition 5.4 If T is a tiling with protoset P, we say that P admits T.

The word 'tile' is used both as noun and as a verb, as in 'any square is a tile' or 'the regular pentagon does not tile the plane'. In either case, when we say that a set of shapes tile, or are tiles, it is understood that the set of tilings they admit is nonempty, and that an infinite number of copies of the prototiles are on hand for constructing the tiling.

It is also reasonable (for the purposes of crystallography) to assume that shapes of the prototiles are rather tame. Definition 5.2 allows tiles of unbounded size, tiles with holes in them, or even – if we do not require a finite protoset – tilings in which infinitely many tiles meet at a common vertex (for examples of tiles and tilings with these and other unusual properties see G&S). So we will place some restrictions on the shapes of the tiles and the way they fit together. Specifically, we will assume that, for a fixed protoset P of a tiling $T \subset E^n$,

 (i) each prototile is homeomorphic to (i.e. can be continuously deformed into) an n-dimensional ball;
 (ii) the intersection of any pair of tiles $T_i \cap T_j$ is a connected set (the empty set is connected!).

We do not require the tiles to be convex or even to be polytopes, although both will be the case for most of the tilings we consider. When the tiles are n-dimensional polytopes, we will use the term *facet* for their $(n-1)$-dimensional faces. Faces of lower dimension will be specified by their dimension except that, as usual, one-dimensional faces will be called edges and zero-dimensional faces will be called vertices.

When the tiles of a tiling share whole facets the tiling is called, not surprisingly, *facet to facet*. When a tiling is not facet to facet , it is necessary to distinguish between the vertices (and the other k-faces) of a tile and of the tiling.

Definition 5.5 A k-face of a tiling is an intersection of $n - k + 1$ tiles of dimension k that is not contained in a face of higher dimension.

Thus a facet ($k = n - 1$) of a tiling is an $(n - 1)$ dimensional intersection of two tiles, and a vertex of a tiling is any intersection of $n + 1$ or more

tiles that consists of a single point. According to this definition, the square
tiles in Figure 5.12(b) have four, five, or six vertices! As another example,
four cubes in E^3 can share an edge; the interior points of the edge are not
vertices.

From now on we will assume that the protoset P is finite. It follows that
the tilings we consider are

(i) normal: for each tiling there are positive real numbers r_0 and R_0
such that every tile contains a ball of radius r_0 and every tile is
contained in a ball of radius R_0;

(ii) locally finite: every ball of finite radius in E^n meets only finitely
many tiles.

Property (i) always holds when the protoset is finite. To show that (ii)
holds as well, we prove

Proposition 5.1 Normal tilings are locally finite.

Proof Suppose that a tiling is normal but not locally finite. Let $\overline{B}_x(a)$ be a
ball of radius a which meets infinitely many tiles. Now expand the radius
of this ball by $2R_0$ so that all the tiles meeting the original ball are
contained in it. Since a ball of radius r_0 can be inscribed in each of the tiles
and the tiles do not overlap, $\overline{B}_x(a + 2R_0)$ contains infinitely many balls of
the same nonzero radius, which is impossible.

The converse is false in general, but it is trivially true when the protoset
is finite.

Normality is the analogue, for tilings, of the Delone condition for point
sets (Definition 1.5); property (ii) is the analogue of the local finiteness
property for point sets described in Proposition 1.1.

5.3 Atlases

In Chapter 1 we defined r-atlases for r-stars of point sets. Now we will
extend this idea to tilings, but we will need the more general idea of a
patch. Let T be a tiling of E^n and O any connected, bounded subset of E^n.

Definition 5.6 The patch of T determined by O is the set of all tiles of T
whose intersection with O is nonempty, together with any tiles needed to
make the union of the tiles of the set homeomorphic to a disc.

When O is a ball $\overline{B}_x(r)$ we will speak of 'a patch of radius r'. Figure 5.4

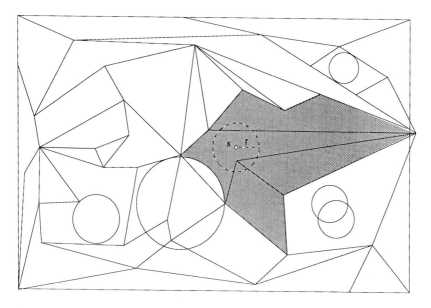

Fig. 5.4 The patch determined by $\overline{B}_x(r)$.

shows such a patch; notice that neither the center nor the radius of the ball is uniquely determined by the patch.

We also need the concept of *corona*. Coronas are usually – but not always – patches.

Definition 5.7 Let T be a tile in a tiling $\mathcal{T}$. The (first) corona of T, $C_1(T)$, is the set of all tiles in $\mathcal{T}$ that meet T:

$$C_1(T) = \{T' \in \mathcal{T} | T' \cap T \neq \emptyset\}.$$

Definition 5.8 Let v be a vertex in a tiling $\mathcal{T}$. The (first) vertex star at T, $W_1(v)$, is the set of all tiles in $\mathcal{T}$ that meet v:

$$W_1(v) = \{T' \in \mathcal{T} | T' \cap v \neq \emptyset\}.$$

Figure 5.5 shows a corona and a vertex star in the tiling of Figure 5.1.

We can also define the second corona, $C_2(T)$, to be the set of tiles in $\mathcal{T}$ that meet $C_1(T)$, and so forth; similarly, the second vertex star $W_2(v)$ is the set of tiles in $\mathcal{T}$ that meet $W_1(v)$.

In most of the tilings of interest to us, there are only a finitely many different kth coronas, and only a finite number of different kth vertex

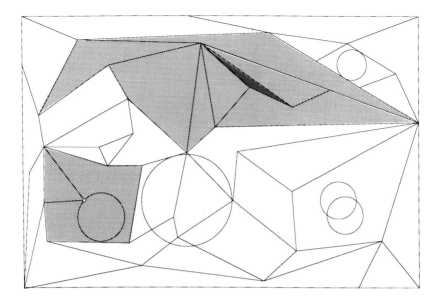

Fig. 5.5 A first corona and a vertex star.

stars. (As before, what 'different' means depends on the rule $\mathcal{M}$ of identifications.) It is useful to think of the set of all patches of a certain type (e.g. coronas or vertex stars) and radius as an *atlas* that contains the local configurations from which the tiling is built (Figure 5.6). In Chapter 7 we will see that some families of tilings are completely characterized by atlases of finite radius, while others are not.

Definition 5.9 The set of the different first coronas that occur in a tiling $\mathcal{T}$ will be called the (first) corona atlas of $\mathcal{T}$. The atlas of all (first) vertex stars that occur in $\mathcal{T}$ is its vertex atlas.

One of the most important properties of normal tilings is the 'normality lemma' of G&S (p. 125). Let $s > 0$, let $\mathcal{T}$ be a tiling, and consider the patches determined by the concentric balls $\overline{B}_x(r)$ and $\overline{B}_x(r + s)$. If $\mathcal{T}$ is normal, then as the radius r increases (but with s staying fixed), the number of tiles that meet the annular region bounded by $\overline{B}_x(r)$ and $\overline{B}_x(r + s)$ will be an increasingly small fraction of the total number of tiles in the patch determined by $\overline{B}_x(r)$. To state this result precisely (though without proof – for the proof, see G&S), let $t_x(r)$ be the number of tiles in the patch $\overline{B}_x(r)$.

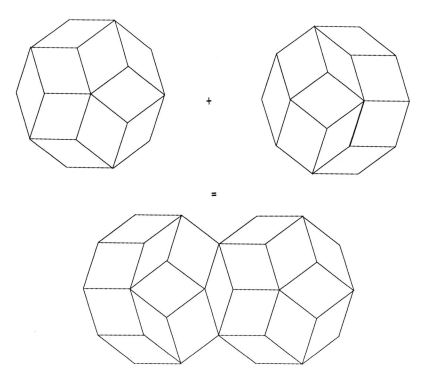

Fig. 5.6 The tiling assembles itself from copies of the configurations of its r-atlases.

Theorem 5.1 If $\mathcal{T}$ is a normal tiling, then for every $s > 0$,

$$\lim_{r \to \infty} \frac{t_x(r + s) - t_x(r)}{t_x(r)} = 0.$$

We have, immediately, an analogue for a Delone set Λ: let $p_x(r)$ be the number of points of Λ in the ball $\overline{B}_x(r)$. Then

Corollary 5.1 For every $s > 0$,

$$\lim_{r \to \infty} \frac{p_x(r + s) - p_x(r)}{p_x(r)} = 0.$$

Theorem 5.1 and its corollary suggest that 'finite size effects' can be safely ignored for sufficiently large patches of tilings or Delone sets. We will use this implicitly in Chapter 8 when we infer properties of the diffraction patterns of infinite tilings from the patterns produced by patches.

5.4 Which shapes are tiles?

Which shapes can serve, singly or together with other shapes, as prototiles for a tiling? This question has been shown to be undecidable! No algorithm exists for determining whether a given shape will tile the plane. Remarkably, the existence of such an algorithm is equivalent to the nonexistence of a protoset that tiles *only* nonperiodically. Thus undecidability was established when Berger (1966) found a set of 20 426 squares with colored edges (called 'Wang tiles'), the colors constituting 'matching rules' that forced nonperiodicity. The connection between Wang tiles and the nonalgorithm is described in detail by G&S, who also give several examples of much smaller sets of Wang tiles.

Since there is no algorithm, we have to use special methods to test particular shapes for tileability. The first test that comes to mind is trial and error, but even as great a thinker as Aristotle stumbled in a very simple case (see Chapter 1). One trusts that he would have discovered his error if he had constructed models, but even a reasonable amount of experimentation with models may be inconclusive in more subtle cases. There are shapes that can fill a bounded region but are not tiles because the finite configuration cannot be continued to fill all space. Figure 5.7 shows a single polygon with this property.

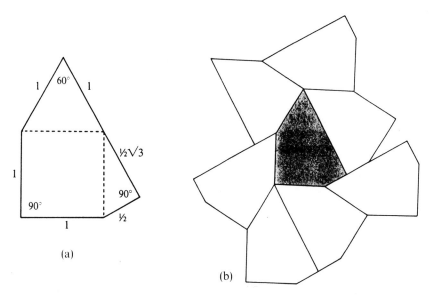

Fig. 5.7 (a) The prototile. (b) The first corona. This configuration cannot be continued (From G&S.)

This prompts a question known as *Heesch's problem*: *for which positive integers does there exist a single shape for which we can construct the first k but not the* $(k + 1)$*st coronas?*

The number k is called the *Heesch number* of the shape. The polygon in Figure 5.7 has Heesch number 1. Several polygons with Heesch number 2 are known (Fontaine, 1991) but only one example with Heesch number 3 (Figure 5.8) has been found so far; it was discovered by Robert Ammann.

Heesch's problem addresses the growth of configurations, a problem that arises in many contexts. For example, it is thought that crystallization begins with the formation of a 'nucleus' or cluster of atoms; as the cluster continues to grow, it may become unstable and disperse, or it may become stable and grow into a crystal. Is there a critical threshold for the size, such that if the cluster manages to attain that size its future as a crystal is assured? The analogue, for tilings, would be the existence of an upper bound for Heesch numbers. G&S conjecture that a bound exists, but the question is still open.

Heesch's problem does not address the question of whether a given set of prototiles might tile an infinite space if their copies were juxtaposed in some other way. Often, however, we are concerned less with the question of whether a particular configuration can be extended than whether its prototiles can tile in *some* way. The following fundamental result is also stated here without proof, since the proof is given in G&S. Although G&S restricted their attention to tilings of the plane, their proof is valid in E^n.

Theorem 5.2 (the extension theorem) Let P be any finite set of shapes in E^n. If, for all $r > 0$, copies of these shapes can be used to construct a patch of radius r, then P admits a tiling of E^n.

Unlike the configurations in Heesch's problem, the finite tilings (patches determined by a sequence of increasingly large balls) need not be successive extensions of the same patch. Remarkably, it is not even necessary that any of the finite tilings be extendable (Figure 5.9)! So this theorem tells us only that a tiling with the given protoset *exists*; it does not tell us how to construct it.

5.5 Orderliness

In this section we restate, for tilings, some of the concepts defined for point sets in earlier chapters.

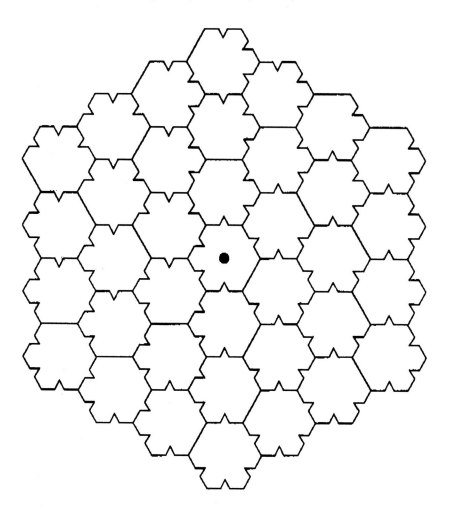

Fig. 5.8 A prototile with Heesch number 3, due to Robert Ammann.

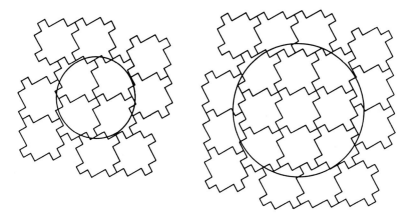

Fig. 5.9 This shape is a tile, by the extension theorem. (From G&S.)

Definition 5.10 A tiling of E^n is periodic if it admits translations in n linearly independent directions.

Thus the symmetry group G of a periodic tiling contains a translation subgroup isomorphic to Z^n. It follows from Theorem 1.1 that the tiling is a union of a finite number of equivalence classes of tiles, the tiles in each class being related by the translations of a lattice.

Every quadrilateral, convex or not – as long as it does not intersect itself – is the prototile for a periodic tiling of the plane; the following proof of this statement could – and should – be taught in elementary school. Let A, B, C, and D be the vertices of a quadrilateral. We generate new copies of the quadrilateral by rotating it 180° about the midpoints of its edges. This process ensures that all four vertex angles will appear at each vertex of the tiling. Since the sum of the angles of a quadrilateral is 360°, the tiles fit together without gaps or overlaps (Figure 5.10). We can continue this rotation process to tile the entire plane.

Each tile is related to its adjacents by rotation, and tiles of the same orientation are related by translation. In other words, the symmetry group G of the tiling has the coset decomposition $G = T + rT$ where $T \simeq Z^2$ is the translation subgroup and r is a rotation of order 2.

Every periodic tiling can be partitioned into unit cells; Figure 5.11 shows one way to do this for the tiling by quadrilaterals that we have just constructed.

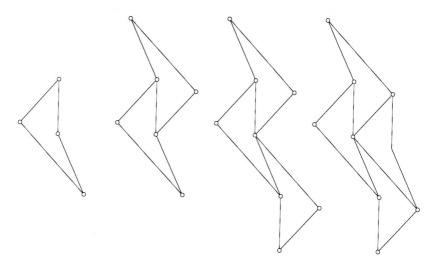

Fig. 5.10 Four copies of any quadrilateral fit together at a vertex.

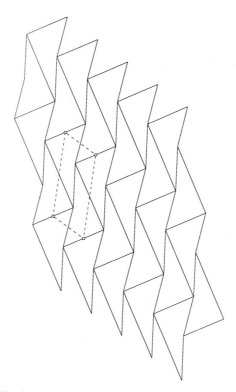

Fig. 5.11 The tiling is periodic; the parallelogram (dashed lines) is a unit cell.

The definition of a periodic tiling is consistent with Definition 1.9, as is the next one:

Definition 5.11 A nonperiodic tiling is one that admits no translations. Tilings that admit translations in k linearly independent directions, where $1 < k < n$, are subperiodic.

The tiling in Figure 5.12(a) is nonperiodic because the point from which the spiral originates is unique and thus there can be no translational symmetry. The tiling in Figure 5.12(b) is subperiodic. Many protosets admit both periodic and nonperiodic tilings (Figure 5.13).

The next definition is the analogue, for tilings, of Definition 1.8.

Definition 5.12 A tiling T is said to be repetitive if every bounded configuration of tiles is relatively dense in the tiling, that is, if for every patch of T there exists a $c > 0$ such that every ball of radius c contains a copy of the patch.

Peroidic tilings are always repetitive, but it is easy to find examples of nonperiodic or subperiodic tilings that are not (Figure 5.12).

A key subject of quasicrystal geometry is the class of nonperiodic tilings whose nonperiodicity is *forced* by matching rules that govern the ways in which the tiles, or patches of tiles, can be put together. Matching rules are thought to be an important constituent of any model for real crystals, insofar as they are analogues of the forces governing the self-assembly of real materials from their constituent atoms and molecules. A model of growth in which nonperiodicity requires intelligent intervention at any stage is not likely to correspond to a physical process.

Thus it is important to distinguish between protosets that admit both periodic and nonperiodic tilings, and those that admit *only* nonperiodic tilings.

Definition 5.13 A protoset is said to be aperiodic if it admits only nonperiodic tilings. Tilings with aperiodic protosets are called aperiodic tilings.

The most famous aperiodic protosets are the several sets of polygons discovered by Penrose; Chapter 6 is devoted entirely to them.

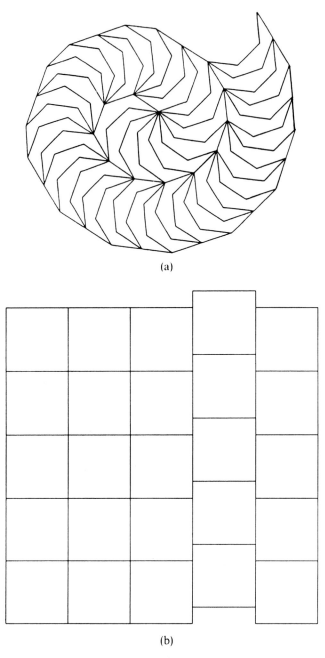

(a)

(b)

Fig. 5.12 (a) The spiral tiling is nonperiodic. (b) The shifted square tiling is subperiodic.

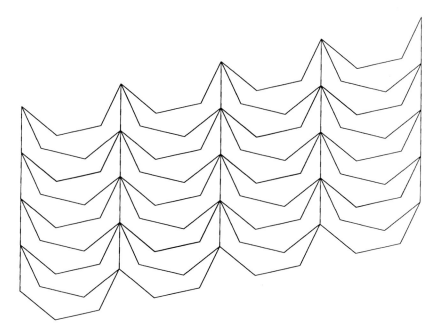

Fig. 5.13 The prototile of the spiral tiling also admits a periodic tiling of the plane.

In most cases it is important to distinguish between the tile shapes, which may be very simple and quite capable of serving as the prototiles of a periodic tiling, and the aperiodic prototiles, which are the same shapes suitably decorated and equipped with 'matching rules' that force nonperiodicity. For example, the shapes of the Wang tiles, mentioned above, are identical squares. But if we color the edges, as prescribed by one or another set of rules, and insist that the colors of the common edge of adjacent tiles must match, then only nonperiodic tilings can be constructed.

Definition 5.14 A set of matching rules for a protoset P is a finite, minimal atlas of the local configurations that may appear in the tilings admitted by P. (An atlas is minimal if none of its configurations is implied by other elements of the atlas.)

Usually the atlas consists of allowed facet pairings, but sometimes it is more convenient (or even necessary) to use kth corona or vertex atlases, for some finite k.

Matching rules can also be expressed as a written list; usually they can be built into the prototiles by marking them in some way or by deforming their facets (see, for example, Figure 5.3 and Figure 6.5).

Notice that Definition 5.14 does not require the rules to force nonperiodicity. Although matching rules are usually discussed in the context of aperiodicity, it is important to note that they are useful for understanding periodic tilings, too. Indeed, the enumeration of the types of isohedral tilings of the plane (Grünbaum and Shephard, 1977; see also G&S) was carried out by the systematic analysis of symbols that recorded the symmetry of polygonal tiles and their orientations with respect to each of their adjacents; the results were presented in visual form as tilings whose edges carried this information implicitly.

Examples of matching rules for aperiodic protosets, and the general problem of finding matching rules, will be discussed in Chapter 7.

Finally, we define local isomorphism for families of tilings:

Definition 5.15 Two tilings T_1 and T_2 belong to the same local isomorphism class if every bounded configuration that appears in one of them also appears in the other.

It follows that if T_1 is repetitive, so is T_2.

5.6 Some construction techniques

In this section we describe several techniques for constructing nonperiodic tilings of E^n. Some of the construction methods are obvious generalizations of the methods for producing one-dimensional nonperiodic Delone sets introduced in Chapter 4 (but there is no higher-dimensional analogue of the circle map). In the one-dimensional case, every Delone set defines a tiling in a natural way: the tiles are simply the intervals between points. But when $n > 1$, finding prototiles that fit together to fill space can be very difficult. We will discuss particular examples (the Penrose tilings and many others) in subsequent chapters; here we concentrate on general methods for producing them. Some readers may prefer to go directly to Chapter 6, returning to these more general considerations after studying that example in detail, although much of this material is assumed in that chapter. Perhaps the best approach is to read these chapters in the order in which they were written, that is, going back and forth between them.

5.6.1 *Voronoï decompositions*

Every Delone set $\Lambda \subset E^n$ induces an associated 'natural' tiling $\mathcal{V}(\Lambda)$, the Voronoï tessellation of E^n. When Λ is nonperiodic, so is $\mathcal{V}(\Lambda)$.

To construct $\mathcal{V}(\Lambda)$, construct the Voronoï cell of each point of Λ as explained in Chapter 2 (more efficient construction algorithms exist, but that does not concern us here). The tiles (the Voronoï cells of the points of Λ) are convex and share whole facets. No two Voronoï cells of a generic Delone set will be alike, but the Voronoï tessellation will always be normal.

Proposition 5.2 The Voronoï tessellation induced by a Delone set $\Lambda \subset E^n$ is a normal tiling of E^n.

Proof Let $x \in \Lambda$. Since the minimum distance between any two points of Λ is $2r_0$, and the facets $V(x)$ lie on the bisectors of the lines joining x to its neighbors, the distance from x to the nearest facet of $V(x)$ must be greater than or equal to r_0. Thus a ball of radius r_0 can be inscribed in $V(x)$. Next, let v be any vertex of $V(x)$. By Proposition 2.6, v is equidistant from x and at least n other points of Λ, those that lie at the centers of the Voronoï cells in its vertex star. This means that v is the center of a ball of radius $|v - x|$ containing no points of Λ in its interior, so $|v - x| \leq R_0$, where R_0 is the covering radius of Λ (see Definition 1.4). Thus $V(x)$ is contained in the ball $\overline{B}_x(R_0)$.

Corollary 5.2 To construct $V(x)$ we do not need to join x to all the other points of Λ, but only to those that lie inside a ball $\overline{B}_x(2R_0)$.

The Voronoï tessellations induced by nonperiodic Delone sets have not received much attention, but they are well worth studying. For example, Figure 6.12 shows a portion of the Voronoï tessellation induced by the set of vertices of a Penrose tiling.

5.6.2 *Dualization*

To every tiling there corresponds a family of *dual* tilings. (Do not confuse dual tilings with dual lattices!)

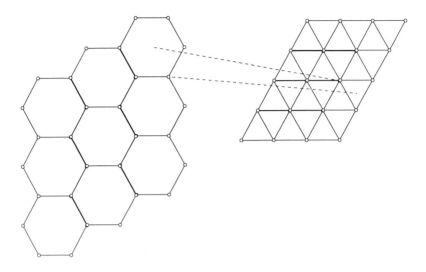

Fig. 5.14 There is a one to one inclusion-preserving map between the *k*-dimensional faces of T and the $(n - k)$-dimensional faces of a dual tiling T'.

Definition 5.16 Two tilings T and T' of E^n are said to be dual if there is a one to one inclusion-reversing map between the *k*-dimensional faces of T and the $(n - k)$-dimensional faces of T', $k = 0, \ldots, n$.

In particular, a vertex *v* of T ($k = 0$) corresponds to a tile T of T' ($k = n$), an edge *e* ($k = 1$) meeting *v* corresponds to an $(n - 1)$-dimensional face of T, and so forth (Figure 5.14).

   This definition of duality describes a purely *combinatorial* relationship between skeletons of vertices, edges, *k*-faces, etc. Such a relationship can be realized in infinitely many ways. For example, Figure 5.15(a) shows a tiling of the plane by hexagons, squares, and triangles, with four edges meeting at every vertex. The tiles of any dual tiling must be quadrilaterals (at least combinatorially) with vertex cycle (6,4,3,4), but beyond these there are no constraints on the tile shapes. Two very different realizations of the dual are shown in Figure 5.15(b) and (c). To convince yourself that (c) is in fact a dual of (a), identify its network of vertices and edges (with the help of Definition 5.5).

Definition   5.17 Two tilings are orthogonally dual if, for $k > 0$, the corresponding *k* and $(n - k)$-faces are orthogonal.

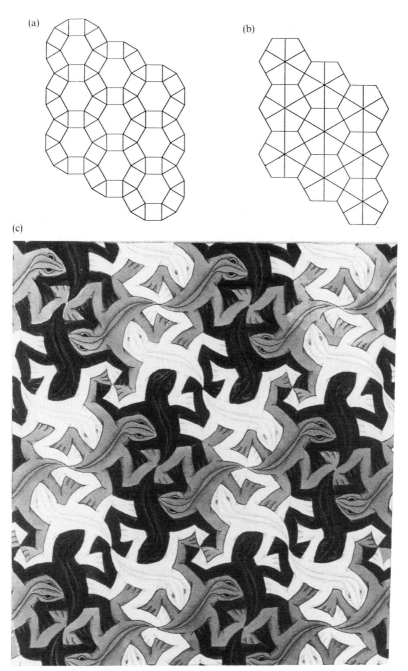

Fig. 5.15 (a) A tiling. (b) A tiling dual to (a). (c) Another tiling dual to (a) (M.C. Escher Foundation – Baarn – Holland. All rights reserved.)

The most familiar examples of dual tilings are those which are both orthogonally dual and *dually situated* (Figure 5.15(a) and (b)).

Definition 5.18 Two tilings are dually situated if the (relative) interior of each k-face of one tiling intersects the (relative) interior of precisely one $(n - k)$-face of the other, and conversely.

As we noted above, a vertex v of a Voronoï tiling lies at the center of a ball containing on its boundary the points of the Delone set Λ belonging to its vertex star $W_1(v)$. These balls, whose interiors are empty (of points of Λ), cover E^n. The convex hulls $\mathrm{Conv}(W_1(v) \cap \Lambda)$ of the points $W_1(v) \cap \Lambda$ are the tiles of a tiling, also induced by Λ, called the Delone tiling; this tiling is orthogonally dual to $\mathcal{V}(\Lambda)$. Like the Voronoï tiling, the Delone tiling is facet to facet and locally finite. In Figure 5.14, the tiling by hexagons is a Voronoï tiling and the other its Delone dual.

Definition 5.19 The Delone tiling induced by the Delone set Λ is the union

$$\bigcup_{v \in \mathcal{V}(\Lambda)} \mathrm{Conv}\big(W_1(v) \cap \Lambda\big).$$

Again, terminology can be misleading. Do not confuse a Delone set Λ with its associated Delone tiling! The Voronoï tessellation induced by Λ is a decomposition of space into tiles whose centers are the points of Λ and whose vertices are the centers of its 'holes' or empty balls, while the tiles of the Delone tiling are centered at those vertices and have the points of Λ as *their* vertices. These dual tilings give us complementary, and very useful, ways of looking at the structure of Λ. Delone was the first to exploit this duality in a general way, although Voronoï had noted it earlier.

5.6.3 Substitution

Definition 5.20 A tiling is said to be *hierarchical* if its tiles can be merged (composed) to form a tiling on a larger scale with a finite protoset, and these tiles can then be composed to form a tiling on a still larger scale with a finite protoset, and so on *ad infinitum*.

Hierarchical structures of many kinds (not all of which can be interpreted as tilings) are receiving considerable attention in many areas of science today; see, for example, (Aksay *et al.*, 1992).

We will restrict our attention to the special case when the prototiles of the larger-scale tiling are similar to (i.e. rescaled versions of) the smaller ones. When the composition matrix is primitive (see Chapter 4 or Appendix I) such a tiling is called a *substitution tiling*. This terminology suggests a close relation both to substitution sequences and to substitution dynamical systems (relations do exist and are being studied; see e.g. Radin (1992) and Robinson (1993)).

As a simple example of a two-dimensional substitution tiling, consider the 'chair' prototile made of three squares. The chair can be decomposed into four congruent chairs, then each of these chairs can be so decomposed, and so on (Figure 5.16). If we enlarge the chairs after decomposition until the small tiles reach the sizes of the original, and iterate this two-step procedure *ad infinitum*, then in the limit we will obtain a substitution tiling of the plane.

Let T be a substitution tiling. We will say that the tiles of the original tiling T are *tiles of level 0*, the tiles at the first hierarchical level are *tiles of level 1*, and so forth.

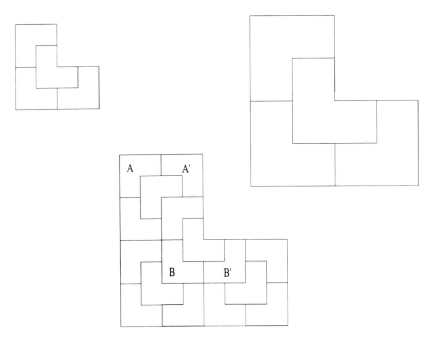

Fig. 5.16 The 'chair' can be decomposed into four chairs similar to itself. There are four prototiles in the infinite tiling; they are congruent but differ in orientation.

Let $T_1, T_2, \ldots, T_k$ be the prototiles of $\mathcal{T}$, and let $T_1', T_2', \ldots, T_k'$ be the corresponding tiles at level 1. Assuming that each T_j' is a union of a whole number of 0-level tiles, we can associate to the composition rule a linear map $\mathcal{U}$:

$$
\begin{aligned}
T_1' &= u_{11} T_1 + u_{21} T_2 + \cdots + u_{k1} T_k \\
T_2' &= u_{12} T_1 + u_{22} T_2 + \cdots + u_{k2} T_k \\
&\;\;\vdots \\
T_k' &= u_{1k} T_1 + u_{2k} T_2 + \cdots + u_{kk} T_k
\end{aligned}
\tag{5.1}
$$

where the coefficients u_{ij} are nonnegative whole numbers that specify the number of copies of T_i in T_j'.

We will always assume that the integer matrix $\mathcal{U} = (u_{ij})$ is primitive.

A word of warning: in principle, any primitive matrix can serve as the linear map $\mathcal{U}$ of a substitution tiling; the difficulty lies in finding a protoset that obeys it. There does not appear to be any general method for constructing sets of shapes that can be decomposed into smaller copies of themselves. And even if we do have a set of prototiles that can be grouped into larger tiles according the rule $\mathcal{U}$, $\mathcal{U}$ only tells us how many copies of each 0-level prototile occur in each tile of level 1, not how the smaller tiles are arranged in the larger ones. In fact, this can sometimes be done in more than one way (for an instructive example, see Godrèche, Luck, Janner and Janssen, 1993). And even when there is only one way to put the tiles together, the geometry of the decomposition can be rather complicated to describe. For example, although the volumes of the tiles T_j' are always linear sums:

$$
\mathrm{vol}(T_j') = \sum_{i=1}^{k} u_{ij} \, \mathrm{vol}(T_j),
$$

'one' copy of a prototile may in fact appear in the larger tile as two half-tiles (see, for example, Figures 6.13 and 7.4.)

From now on, we will implicitly assume that $\mathcal{U}$ comes with instructions that select one particular way to build $T_1', T_2', \ldots, T_k'$ from the protoset.

Just as in the one-dimensional case (Chapter 4), we can use $\mathcal{U}$ to compute the relative volumes of the prototiles. Suppose that the volumes of $T_1, \ldots, T_k$ are $\alpha_1, \ldots, \alpha_k$. Suppose also that the volumes of the composed level-1 prototiles T_j' are $\lambda \alpha_j$, $j = 1, \ldots, k$, where $\lambda > 1$. Then since $\lambda \alpha_j = u_{1j} \alpha_1 + u_{2j} \alpha_2 + \cdots u_{kj} \alpha_k$, we have

$$
\mathcal{U} \vec{\alpha} = \lambda \vec{\alpha},
\tag{5.2}
$$

where $\vec{\alpha} = (\alpha_1, \ldots, \alpha_k)$. Thus λ must be an eigenvalue of $\mathcal{U}$ and $\vec{\alpha}$ the corresponding right eigenvector.

Starting with some patch and iterating $\mathcal{U}$, we obtain successively larger numbers of tiles. We can define population vectors $\vec{p}_m$ for the tiles of the patches, just as we did in the one-dimensional case, and we can compute the relative frequencies of the prototiles by studying the behavior of the population vectors under iteration. Let $\vec{v}_1, \vec{v}_2, \ldots, \vec{v}_k$ be unit eigenvectors of $\mathcal{U}$, corresponding to eigenvalues $\lambda_1, \ldots, \lambda_k$, where λ_1 is the leading eigenvalue. Then, writing $\vec{p}_0$ as a linear combination of the $\vec{v}_j$s, we have

$$\vec{p}_0 = \beta_1 \vec{v}_1 + \cdots + \beta_k \vec{v}_k;$$

$$\left(\frac{1}{\lambda_1}\right)^m \vec{p}_m = \beta_1 \vec{v}_1 + \beta_2 \left(\frac{\lambda_2}{\lambda_1}\right)^m \vec{v}_2 + \cdots + \beta_k \left(\frac{\lambda_k}{\lambda_1}\right)^m \vec{v}_k. \qquad (5.3)$$

Since $\mathcal{U}$ is primitive, $\lambda_1 > |\lambda_j|, j = 2, \ldots, k$ and so $(\lambda_j/\lambda_1)^m \to 0$ as $m \to \infty$. Thus the slope of the population vector approaches that of $\vec{v}_1$, which means that the relative frequencies of the prototiles are proportional to the components of the left eigenvector belonging to λ_1.

Definition 5.21 A hierarchical tiling $\mathcal{T}$ is said to be *uniquely* hierarchical if its level-m tiles can be composed into level-$(m+1)$ tiles in only one way, for all $m = 0, 1, 2, \ldots$

Note that Definition 5.21 is concerned with a particular tiling $\mathcal{T}$, not with any other tilings that the prototiles might admit. Note also that not every hierarchical tiling is uniquely hierarchical. For example, in the ordinary tiling by squares there are many ways to group the squares into larger ones.

A uniquely hierarchical tiling $\mathcal{T}$ has a 'substitution atlas', the set $\mathcal{A}_1$ of level-1 prototiles. From it we can make a list, for each level-0 prototile, of the level-1 prototiles into which it can be embedded, with unambiguous instructions for doing so. Using the atlas $\mathcal{A}_1$ we can build many different hierarchical tilings (in the same local isomorphism class as $\mathcal{T}$), sometimes uncountably many. There are many variations on this construction, and they go by many different names – inflation, deflation, substitution, composition, decomposition, and so forth. There is no standard terminology: any one of these names may signify one process to one author and another to another. We will adopt the following terminology.

1. To generate a tiling by *decomposition*, we choose any tile at level 0, say T_1, and decompose it into smaller copies of the prototiles $T_1, \ldots, T_k$ as

if it were T_1'; then we rescale these small tiles (level-(-1)?) to the sizes of the originals. Next, we decompose each of the tiles in the patch we have just constructed, rescale, decompose again, and so forth *ad infinitum*. Proofs that this leads to a tiling of E^n can be found in Lunnon and Pleasants (1987a) and elsewhere; it does *not* follow from the extension theorem!

2. Alternatively, we can think of the process as *substitution*: starting with one 0-level copy of T_1, we substitute the appropriate patch from $\mathcal{A}_1$ for it, then substitute appropriate patches for each of the tiles in the first patch, and so forth. This way of looking at the process avoids the explicit rescaling of the decomposition method, but leads to the same family of hierarchical tilings.

3. We can also generate the same classes of tilings by a method called *up–down generation* (de Bruijn 1989a, 1990). We begin with any prototile, say T_1, and any level-1 prototile T_j' in which it appears, that is, for which $u_{1j} \neq 0$ (we assume that at least one such coefficient exists). We then build T_j' with copies of 0-level tiles. Next choose any prototile at level 2 in which T_j' appears, and construct it using tiles of level 1; then subdivide the level-1 tiles into tiles of level 0. Repeating these steps *ad infinitum*, we obtain a tiling of E^n.

Up–down generation has an elegant description as an infinite path through a finite directed graph (a finite state automaton). Each prototile is represented as a node of the graph, and the node representing T_i is connected by a directed edge to the node representing T_j if and only if T_i can be embedded in T_j' (Figure 5.17). There is a one to one correspondence between the tilings constructed by up–down generation and families of infinite paths through the graph. We will discuss up–down generation in more detail in Chapter 6.

Substitution tilings are sometimes said to be *self-similar*, but this term is a little misleading: the tiling at level $k + 1$ need *not* be a (rescaled) copy of the tiling at level k, even though all of its prototiles are, and even though the two tilings may be locally isomorphic. For example, we will see in Chapter 6 that a level-$(m + 1)$ Penrose tiling is, except in certain exceptional cases, different from the level-m tiling that gave rise to it, and we will use this fact to show that the number of different Penrose tilings is uncountable.

How can we know whether these or any substitution tiling is nonperiodic? Certainly we cannot find out by inspecting any finite portion of the tiling. But there are at least two simple tests:

• If the ratio of any two components of a left eigenvector belonging to the leading eigenvalue of $\mathcal{U}$ is irrational, then the tiling is necessarily

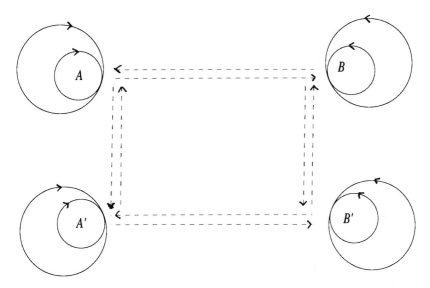

Fig. 5.17 The directed graph for up – down generation of chair tilings.

nonperiodic (for reasons explained in Chapter 4). Also:

- If T is uniquely hierarchical, then it is nonperiodic.

This second test is particularly useful in cases where the first does not apply.

Theorem 5.3 A uniquely hierarchical tiling is nonperiodic.

Proof The main idea is that any translation that carries a uniquely hierarchical tiling onto itself must do this simultaneously at every hierarchical level. This eventually leads to a contradiction: a translation is a shift through a fixed, finite distance, but the tile sizes at levels $0, 1, \ldots k, k + 1, \ldots$ increase without bound.

Since our protoset is finite, we can inscribe a circle of radius c in the 0-level tiles, a circle of radius λc in the level-1 tiles, and so on (where, again, λ is an eigenvalue of the substitution matrix). After k compositions the radius of the incircle will be $\lambda^k c$. If the tiling had translational symmetry, we would be able to bring it into self-coincidence by a shift through some vector $\vec{u}$ of fixed finite length u. But this would be possible only when $u > \lambda^k c$; otherwise the shift would cause the tiles to overlap their original positions. Clearly, at some level of the hierarchy the incircles will be so large that $\lambda^k c > u$, and then the translation vector will lie entirely inside the incircle of the level-k tiles. Thus translation by $\vec{u}$ is not a symmetry operation for the (sufficiently large) kth hierarchical level of the tiling. It is not a symmetry operation for the original tiling either: if it were, the overlap portions of the composed tiling would contain elementary tiles that belong to more than one composed tile, contradicting the uniqueness of composition. Thus the tiling is nonperiodic.   ❏

For a given primitive substitution matrix $\mathcal{U}$ and protoset P, all of the tilings will belong to a single local isomorphism class.

Theorem 5.4 The tilings admitted by $\mathcal{U}$ and P belong to a single local isomorphism class.

Proof Since $\mathcal{U}$ is primitive, there is an integer $j_1 > 0$ such that the entries of $\mathcal{U}^{j_1}$ are positive. This means that a copy of every prototile appears in *every tiling* at the j_1th hierarchical level. Every finite configuration in the tiling can be obtained from any prototile by sufficiently many substitutions, and thus there are hierarchical levels $j_2, j_3, \ldots$ such that every configuration of suitable size appears in every tiling at those levels. Thus all the tilings produced in this way belong to a single local isomorphism class.   ❏

Corollary 5.3 Any substitution tiling with primitive substitution matrix $\mathcal{U}$ is repetitive.

5.6.4 Multigrids

In 1981 de Bruijn introduced a powerful method for constructing the Penrose tilings of the plane. The method, now usually called the multigrid method, can easily be generalized to construct large families of nonperiodic tilings in E^n with 'parallelepiped' prototiles.

The grid method is inspired by the observation that any edge-to-edge tiling of the plane by rhombs is a 'weave of ribbons' made of rhombs linked by parallel edges (Figure 5.18). Every rhomb has two pairs of parallel edges, say $\{a, a'\}$ and $\{b, b'\}$. Each of the rhombs adjacent to a given rhomb along edges a and a' has another edge parallel to them, as do their adjacents, and so on; the ribbon determined by $\{a, a'\}$ is an infinite chain of adjacent rhombs.

Replacing the ribbons by straight lines orthogonal to the edges that determine them, we obtain a configuration of lines that is a superposition

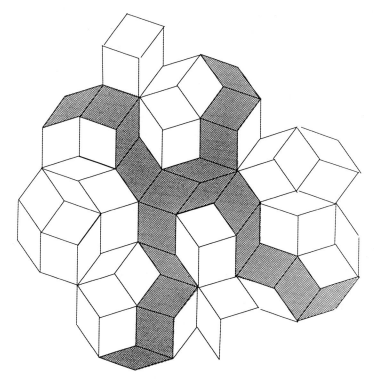

Fig. 5.18 Any edge-to-edge tiling of the plane by rhombs is a 'weave of ribbons' of rhombs linked by parallel edges; two ribbons are indicated here.

of k infinite families of parallel lines, where k is the number of directions of edges in the rhomb tiling. When this configuration is orthogonally dual to the tiling by rhombs we can reconstruct the rhomb tiling from it.

Similarly, many facet to facet tilings of E^n by zonotopes (centrosymmetric polytopes with centrosymmetric facets) are dual to configurations of intersecting families of parallel hyperplanes.

In certain special cases, such as the Penrose rhomb tilings, the line configuration can be realized as the superposition of families of equally spaced parallel lines. Then each family is called a *grid* and the configuration of k grids is called a *multigrid*.

Definition 5.22 A grid in E^n is an infinite family of parallel hyperplanes with fixed interplanar spacing d; a vector of length d orthogonal to the hyperplanes is called a grid vector. A multigrid or, more specifically, a k-

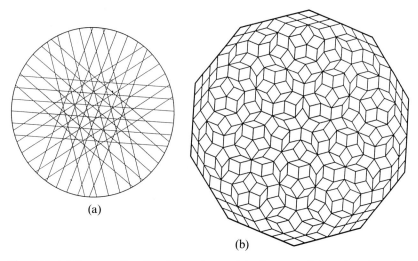

Fig. 5.19 (a) Representing the ribbons by straight lines parallel to the edges that determine them, we obtain a configuration of intersecting families of parallel lines. (b) We can reconstruct the rhomb tiling by dualizing the line configuration. (J. Richter-Gebert; for the postscript code, see Chapter 8.)

grid, is a superposition of k grids. A set of k grid vectors $\vec{\varepsilon}_1, \ldots, \vec{\varepsilon}_k$ is the star of the multigrid.

The multigrid construction for tilings of E^n reverses the process described above: we begin with the multigrid and construct the tiling as an orthogonal dual. (This can still be done when the distances between the lines are not the same for all grids; you can use the code in Section 8.4.3 to experiment with different possibilities.)

For example, Figure 5.19(a) shows a portion of a five-grid or *pentagrid* for which the star vectors have the same length and point to the vertices of a regular pentagon. Notice that the edges of the tiles in (b) are orthogonal to the lines of the grid and thus parallel to the vectors of the star. In fact, the tiles are completely determined by the star: the two nonparallel edges of each rhomb are translates of vectors of the star.

A k-grid is itself a tiling, but it is not normal and its protoset is not finite. To avoid confusion we will use other terminology for its tiles: following de Bruijn, we will call them *meshes.*

The multigrid construction involves a series of steps:

Step 1. Choose a star of k unit vectors, where $k > n$.

Step 2. Superimpose k grids whose hyperplanes are orthogonal to the star vectors. If more than n hyperplanes meet at any point, shift sufficiently many grids away from the origin (in the direction of their star vectors) in such a way that this does not occur.

Step 3. Using the grid star, construct the prototiles of the tiling: the prototile corresponding to a given vertex of the k-grid is the parallelotope spanned by the star vectors corresponding to the hyperplanes that meet there. There will be only a finite number of prototiles since there are only $\binom{k}{n}$ intersection configurations (up to translation).

Step 4. With copies of the prototiles and using the k-grid as a blueprint, construct an orthogonal dual of the k-grid.

Proofs that this process does produce a tiling of the entire plane can be found in de Bruijn (1986a) and Beenker (1982).

The reason for the shifts in step 2 is to avoid 'singular' cases. For example, suppose $n = 2$; then the grids are families of equally spaced parallel lines. We want the tiles of the dual tiling to be rhombs, but the tile corresponding to a point where m grid lines meet will be a $2m$-gon, which is a rhomb only if $m = 2$. We can begin the construction with one line of each grid passing through the origin, and then shift the jth grid, $j = 1, \ldots, k$, through some distance γ_j, in the direction of the jth vector of the star so that at most two lines meet at any point.

If $\gamma_j \in Z$ the shift will bring the grid into self-coincidence, so we may assume that $0 \leq \gamma_j < 1$.

Definition 5.23 A shift vector for a k-grid is a k-tuple of real numbers

$$\vec{\gamma} = (\gamma_1, \ldots, \gamma_k), \quad 0 < \gamma_j \leq 1, \, j = 1, \ldots k.$$

If $\vec{\gamma}$ is such that no more than n hyperplanes intersect in any point, both it and the corresponding k-grid are said to be *regular*; all other $\vec{\gamma}$s and k-grids are *singular*.

Different shift vectors may produce different (nonsuperposable) multigrids, and thus different dual tilings.

By the way, it can be shown that

Proposition 5.3 Regular k-grids exist.

(For a proof of Proposition 5.3 for the special case of the Penrose tilings ($n = 2, k = 5$) see de Bruijn (1981a)). In that case the singular shift vectors are a subset of measure zero.

We will assume from now on that the grids we deal with are regular.

That's essentially all there is to the multigrid construction which, as you can see, is quite general. But there is more information to be extracted from the construction. In particular, it allows us to assign coordinates (k-tuples) to the vertices of the tiles. First we index the parallel hyperplanes in each grid by the numbers $Z + 1/2$. The equations of the hyperplanes are thus

$$\vec{x} \cdot \vec{\varepsilon}_j = l_j + \frac{1}{2}, \quad l_j \in Z, \, j = 1, \ldots, k, \tag{5.4}$$

where $\vec{\varepsilon}_j$ is the jth vector of the grid star. Incorporating the shifts into equations (5.4), we have

$$\vec{x} \cdot \vec{\varepsilon}_j = l_j + \frac{1}{2} + \gamma_j, \tag{5.5}$$

where γ_j is the signed length of the shift of the jth grid.

Any two successive hyperplanes of a grid, for example $l_j - \frac{1}{2} + \gamma_j$ and $l_j + \frac{1}{2} + \gamma_j$, constitute a 'slab'; every mesh of the k-grid lies in the intersection of k such slabs. The meshes can thus be labeled by k-tuples of integers $(l_1, \ldots, l_k)$.

However, not every k-tuple of integers corresponds to a mesh because the corresponding intersection of slabs may be empty. If $\vec{x}$ is a point in the interior of a mesh, then it must satisfy

• *The mesh condition*: there are integers $l_j, j = 1, \ldots, k$, such that the k inequalities

$$l_j - \frac{1}{2} + \gamma_j < \vec{x} \cdot \vec{\varepsilon}_j < l_j + \frac{1}{2} + \gamma_j \tag{5.6}$$

have a simultaneous solution.

The k-tuples that do correspond to meshes can be used to label the vertices of the dual tiling; in this way every vertex of the rhomb tiling is assigned coordinates $(l_1, \ldots, l_k)$. The mesh condition tells us exactly which k-tuples these are.

In general, the dual tiling will be nonperiodic. Moreover, it is always the case that

Proposition 5.4 Tilings dual to regular k-grids are repetitive.

5.6.5 *Projection*

The 'projection method' for constructing nonperiodic point sets was introduced in Chapter 2 and used in Chapter 4 to construct one-dimensional crystals. In this chapter we will use it to construct tilings of the plane and higher-dimensional spaces.

There are actually two variations on the projection construction for tilings. Both produce the same tilings but it is useful to distinguish between them. This is usually done linguistically by calling one the projection method and the other the cut method. But beware: the first of these is often called 'cut and project' – an apt description – and it in turn can be formulated in at least two different ways. One of them, which we will follow, is discussed in detail in (Schlottmann, 1993a, 1993b); for the other formulation, see Oguey, Duneau, and Katz (1988). We will postpone discussion of the cut method until we need it, which will be in Chapter 7.

Let us first recall the canonical projection construction. The basic ingredients are:

(i) an n-dimensional lattice $\mathcal{L}$, which we assume is integral;
(ii) a totally irrational d-dimensional subspace $\mathcal{E}$ (i.e. $\mathcal{E} \cap \mathcal{L} = \{0\}$);
(iii) a regular shift vector $\vec{\gamma} \in E^n$ (see Definition 5.23).

Here the k of Definition 5.23 is our n, and the n in that definition is our d. The regularity of $\vec{\gamma}$ ensures that $\mathcal{E} + \vec{\gamma}$ does not contain a j-face of the Voronoï tessellation induced by $\mathcal{L}$ if $0 \leq j < n - d$.

For simplicity, in this section we will assume that $\mathcal{L}$ is the standard lattice $\mathcal{I}_n$, although our discussion applies to any lattice. As we will see in Chapters 6 and 7, many of the best-known aperiodic tilings can be obtained by projection from the standard lattice.

Since the Voronoï cells of $\mathcal{I}_n$ are hypercubes, the window is the orthogonal projection, into E^{n-d}, of an n-dimensional unit cube $\mathcal{Q}_n$. We note for future reference that the orthogonal projection of $\mathcal{Q}_4$ into E^2 is an octagon, the projection of $\mathcal{Q}_4$ into E^3 is a rhombic dodecahedron (Figure 1.14), the projection of $\mathcal{Q}_5$ into E^3 is a rhombic icosahedron (Figure 6.21), and the projection of $\mathcal{Q}_6$ into E^3 is a rhombic triacontahedron (Figure 1.2); for further details see Coxeter (1973).

Let us first review the one-dimensional case that we studied in Chapter 4. There we projected, onto a line ℓ, the points of $\mathcal{I}_2^p$ whose Voronoï cells were cut by ℓ. This automatically gave us a tiling of the line by one-dimensional tiles (line segments), the projections of the facet vectors joining each Voronoï cell (cut by ℓ) to its neighbors.

We want to find an appropriate analogue for tilings in dimensions greater than one. As in the one-dimensional case, the projected lattice points should be the vertices of the tiles. But how do we construct their faces?

In fact, the answer to this question is provided by the multigrid construction, in which the tilings are obtained as duals of multigrids. It can be shown – we will do it explicitly in Chapter 6 – that an *n*-grid is the intersection of the Voronoï tessellation of E^n induced by $\mathcal{I}_n^p$ with a *d*-dimensional plane $\mathcal{E}$. The *n*-tuples of integers that satisfy the mesh condition are precisely the coordinates, in E^n, of the lattice points whose Voronoï cells are cut by $\mathcal{E}$. Thus the multigrid construction has a natural interpretation in a higher-dimensional space, and the 'projection method', for tilings, turns out to be just a higher-dimensional interpretation of dualization.

What, in fact, is the meaning of multigrid dualization from the higher-dimensional point of view? We have seen that the Delone tiling is an orthogonal dual of a Voronoï tiling. By the definition of duality, there is a one to one inclusion-reversing correspondence between the *j*-faces of one of these tilings and the $n-j$-faces of the other. Thus when we cut a Voronoï tiling by $\mathcal{E}$, we select a subset of faces of the Voronoï tiling and at the same time the corresponding subset of faces of the Delone tiling. The regularity of the shift vector $\vec{\gamma}$ ensures that the dimensions of the faces of the Delone tiling that we select in this way do not exceed *d*.

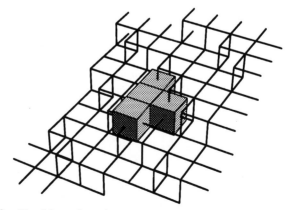

Fig. 5.20 $\mathcal{L} = \mathcal{I}_3$; $\mathcal{E}$ is a plane (not shown). The cubes are some of the Voronoï cells cut by $\mathcal{E}$; the line segments are the faces of the Delone tiling that will be projected onto $\mathcal{E}$.

To construct the tiling of $\mathcal{E}$, we project the selected faces of the Delone tiling onto $\mathcal{E}$. One can show (Schlottmann, 1993a, 1993b) that the projected tiling is the dual of the multigrid $\mathcal{V}(\mathcal{I}_n^p) \cap \mathcal{E}$; indeed, if α is a face of $V(0)$ that is cut by $\mathcal{E}$ and α^* is the corresponding face in the Delone tiling, then $\alpha \cap \mathcal{E}$ corresponds to $\Pi(\alpha^*)$ in the duality.

Since the vertices of the Delone tiling are points of $\mathcal{I}_n^p$, we obtain the same set of projected points as in Chapter 2.

As an example, let $n = 3$ and let $\mathcal{E}$ be a totally irrational plane. Figure 5.20 shows some of the Voronoï cells (cubes) cut by $\mathcal{E}$, together with the corresponding faces of the Delone tiling (line segments).

It is straightforward to compute the local patterns that occur in a projected tiling; in Chapter 6 we will do this explicitly for the Penrose tilings of the plane. The computation is similar to the one-dimensional case, but now it is both more complicated and more interesting.

5.7 Notes

1. The idea of using the term 'aperiodic' for prototiles that admit *only* nonperiodic tilings is due to Grünbaum and Shephard. Although the concept is crucial and there needs to be a word for it, many people who fully understand the distinction persist in using 'nonperiodic' and 'aperiodic' interchangeably, as we do in ordinary speech. The confusion is made worse by the fact that 'aperiodic' is rapidly becoming the preferred way to describe nonperiodic crystals, as we have done in some places in this book. Finding the right word for this phenomenon is another important unsolved problem.

2. For an extensive discussion of the geometry and physics of substitution tilings, see Hof (1992).

3. A single tile that can be decomposed into copies of itself is called a 'rep-tile' (Golomb, 1964).

4. The equivalence of the multigrid and projection methods is implicit in (de Bruijn, 1981a); it is discussed in greater generality and more detail in (Gähler and Rhyner, 1986) and (Korepin, Gähler, and Rhyner, 1988).

5. If you don't see ribbons when you look at Figure 5.18, what do you see? de Bruijn sees stacks; physicists have seen rails, trails, tracks, and rows (de Bruijn, 1987a; Ingersent, 1991). It is tempting to take this as the starting point for a discourse on metaphor and imagery in mathematics, but perhaps this is not the place to do it.

6

Penrose tilings of the plane

Aesthetic delight lies somewhere between boredom and confusion.

(E. H. Gombrich)

This chapter is devoted entirely to the Penrose tilings of the plane and some of their interesting generalizations. The Penrose tilings come in several guises; we will consider mainly the version in which the prototiles are two rhombs, one 'thick' and one 'thin', with arrows on their edges that must be matched when the tiles are put together. The rhombs fit in well with standard crystallographic ideas: they look like two-dimensional versions of the unit cells of periodic crystals, and because they are convex they seem to be more suited as modules for crystal structure than Penrose's other prototiles. But there are mathematical reasons for this choice as well. The rhomb tilings can be constructed by the projection method as well as by substitution; the different constructions give different insights into their remarkable properties. Moreover, there is a three-dimensional analogue of the rhomb tilings, though not – apparently – of the other versions. We will discuss the three-dimensional tilings in Chapter 7.

The remarkable properties of the Penrose tilings have been discovered by various people working independently; the work of Penrose, Conway, and Ammann on the version known as 'kite and dart' tilings is described in G&S . Most of our understanding of the rhomb tilings, however, is due– directly or indirectly – to N. G. de Bruijn. Sections 6.2, 6.3, and 6.4 (and parts of other sections as well) are based on his many papers on this subject. The key reference is (de Bruijn, 1981a); unless otherwise indicated, the reference 'de Bruijn' will always mean that paper.

The best introduction to the Penrose tilings is to play with the tiles. It is not difficult to make your own from cardboard or, even better, plexiglass (or, you can now buy them – see Section 6.7). Be forewarned, however: even if you obey the matching rules, you will probably find, at some stage, that you have constructed an untileable region. When this happens, just remove some tiles and try again. For more about this problem, see below (and also Chapter 7). In Chapter 8 we describe a Mathematica program for drawing Penrose and other substitution tilings.

6.1 Penrose tiles and matching rules

While investigating the Archimedean tilings of the plane in the early 1600s (Kepler, 1619), Kepler discovered the curious, nonrepeating pattern *Aa* (Figure 1.10). He wrote:

> If you really wish to continue the pattern, certain irregularities must be admitted, two decagons must be combined, two sides being removed from each of them ... as it progresses this five-cornered pattern continually introduces something new. The structure is very elaborate and intricate.

Over 350 years later, Penrose (1974) produced a series of three families of aperiodic tilings that would surely have delighted Kepler. In fact, Penrose's first tiling family (Figure 6.1) is essentially a completion of Kepler's *Aa*.

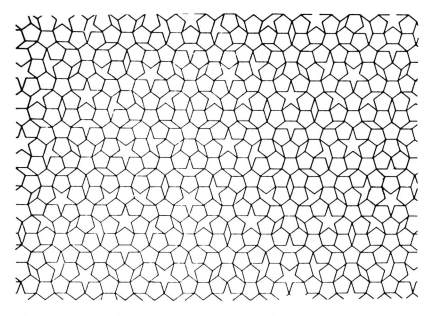

Fig. 6.1 Penrose's first aperiodic tiling. Notice that the prototiles have four different shapes. When the edges are decorated as required by the matching rules, there are six prototiles (see G&S).

This completion was a giant step. In the first place, the Penrose tilings cover the entire plane, while Kepler gave up after tiling a finite region. And while the Penrose tilings are indeed nonperiodic, it does not seem quite right to say that there is 'always something new': not only is the protoset finite, the tiling is repetitive, and all tilings with this protoset belong to a single local isomorphism class. It is likely that Kepler never suspected that 'his' tiling could be generated by substitution (Figure 6.2); this was one of Penrose's key discoveries, from which these two properties follow.

You can still see Kepler's decagons and fused decagons (in all three Penrose families), but now their interiors are decomposed into copies of the prototiles. In the first version, the prototiles have four different shapes: one is a regular convex pentagon and the others are 'gap tiles' that arise as we try to tile the plane with the pentagons. The matching rules require us to mark the tiles or deform the tile edges; the pentagons must be marked or deformed in three different ways (see G&S), so there are actually six prototiles.

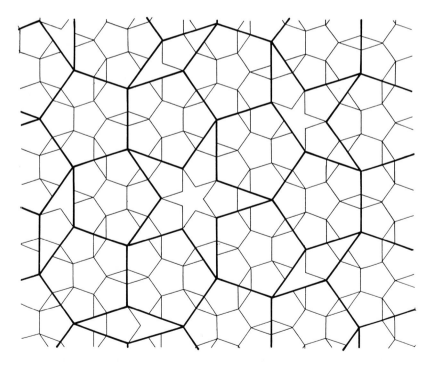

Fig. 6.2 The tiling of Figure 6.1 is a substitution tiling.

Penrose showed that the six prototiles could be replaced by the two quadrilaterals with colored vertices known as kites and darts. Not only are these shapes simpler than the prototiles of the first family, so are their decorations and matching rules: when tiles are juxtaposed along an edge, the colors of the vertices on the edge must match. Notice that the rules forbid juxtaposition either by translation or by rotation about the midpoint of an edge (Figure 6.3). Thus although the prototile shapes are quadrilaterals, we cannot use them to construct the quadrilateral tiling described in Section 5.5.

The rhomb tilings can be derived from the kite and dart tilings by bisecting the tiles into isosceles triangles and merging the triangles into thick and thin rhombs as shown in Figure 6.4; we will denote the rhombs by T and t. Conversely, we can derive the kites and darts from the rhombs; in the terminology of Chapter 7, the kite and dart and the rhomb tilings belong to the same mutually locally derivable family.

The measures of the angles of the T rhombs are 2θ and 3θ and those of the t rhombs are θ and 4θ, where $\theta = \pi/5$. If we assign unit length to the edges, the length of the long diagonal of the T rhomb is $\tau = (1 + \sqrt{5})/2$ and the length of the short diagonal of the t rhomb is $1/\tau$.

Several equivalent versions of the matching rules for the rhombs are shown in Figure 6.5. The edge and vertex decorations in (a) are derived directly from those of the kites and darts. In version (b), only the edges are marked; this is useful because it allows us to regard the tiling as an oriented graph. In (c), the lines (called *Ammann bars*) must continue (as straight lines) across tile boundaries. In (d), the rules are expressed by modifying the shapes of the edges to permit only the allowed juxtapositions.

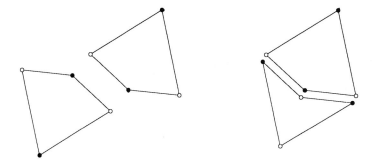

Fig. 6.3 The matching rules for the kite and a dart (the colors of the vertices must match) forbid configurations that can tile by translation.

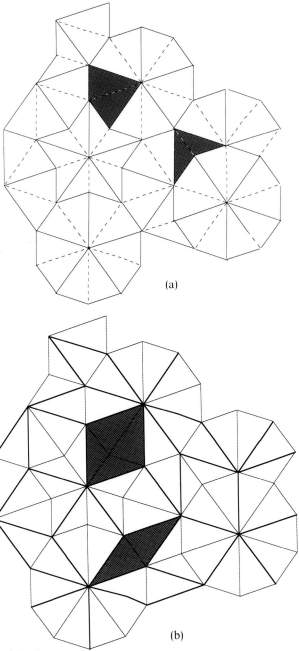

(a)

(b)

Fig. 6.4 Transition from a kite and dart tiling to a rhomb tiling. (a) First bisect the kites and darts (dotted lines). (b) Then merge the triangles into rhombs as shown (heavy lines).

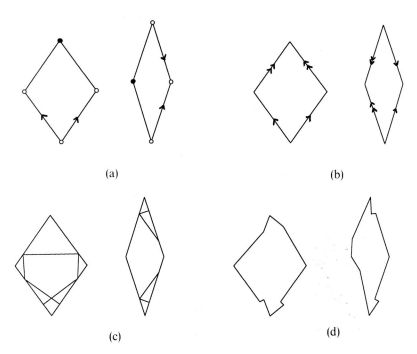

(a)   (b)

(c)   (d)

Fig. 6.5 Four realizations of the matching rules for T and t rhombs. (a) Rules derived from those for the kites and darts. (b) The rules expressed as arrowed edges. (c) The rules expressed as Ammann bars. (d) The rules expressed as edge deformations.

Figure 6.6 shows a large region of a Penrose tiling equipped with single and double arrows; we will always use this version of the rules. (The numbers at the vertices will be explained in Section 6.3.)

It is important to be aware that these matching rules do much more than guarantee nonperiodicity – they force any tiling that obeys them to be a Penrose tiling (and thus a member of a single local isomorphism class). Not all matching rules are so strong. Examples of weaker matching rules will be given in Chapter 7.

As we have already observed, the Penrose rules do not work automatically: some 'locally legal' configurations cannot be extended without breaking the rules. For example, although the matching rules permit us to juxtapose three T rhombs with $3\,\theta$ vertices around a point, Figure 6.7 shows that they do not permit the completion of this configuration even though there is *room* for a t rhomb to be inserted. Similarly, although the juxtaposition of three t rhombs shown below is

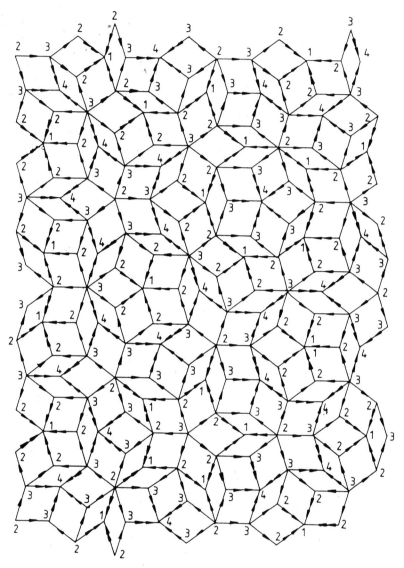

Fig. 6.6 A Penrose tiling, with matching rules expressed as arrows.

also locally legal, this configuration cannot be extended either. *Tilings by T and t rhombs in which these configurations occur are not Penrose tilings!* The central role of the matching rules apparently escapes many authors, since it is not unusual to find incorrect 'Penrose' tilings in the published literature.

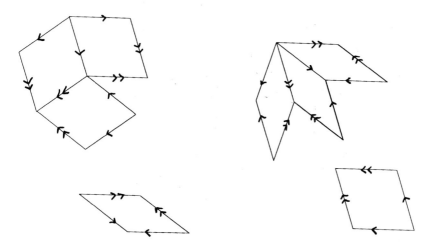

Fig. 6.7 Two incompletable configurations.

A configuration that *can* be continued is said to be 'globally legal'. There are seven globally legal vertex stars in the Penrose tilings, one of which can be constructed with two different matching rule decorations (see Figure 6.8). Each vertex is characterized by a cycle of integers from the set $\{1, 2, 3, 4\}$ denoting the measures of the angles, in multiples of θ, of the rhombs that meet there. If we erase the arrows and the redundant star in Figure 6.8, we have an atlas of unmarked vertex stars. Remarkably, this atlas completely determines the family of Penrose tilings by rhombs.

Theorem 6.1 Any tiling of the plane by T and t rhombs whose stars are restricted to the seven types shown in Figure 6.8 is a Penrose tiling, provided that no two stars that share a rhomb are related by a half-turn about the center of that rhomb.

Proof To prove this theorem, we will show that every tiling satisfying these conditions does in fact obey Penrose's matching rules. First we restate the obvious: each of the seven vertex types can be 'correctly arrowed', one of them – $(2, 2, 2, 2, 2)$ – in two ways. What needs to be proved is that there is never an inconsistency when these stars are extended in accordance with the vertex atlas, and the above proviso.

To show that no inconsistency arises, we take each of the stars in turn and inspect the vertices along its boundary. Each of the boundary vertices is part of an incomplete configuration, and we need to show that if it is completed to a star in the atlas, again with the proviso, there will be no contradiction in the number or direction of the arrows.

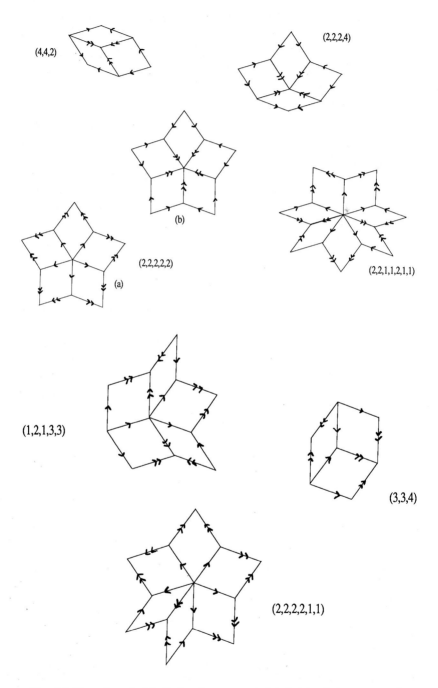

Fig. 6.8 The atlas of globally legal vertex stars labeled by their vertex cycles.

The procedure is completely straightforward except for the case $(2,2,2,2,2)$, which we will discuss in detail below. For example, let us consider vertex type $(2,2,2,4)$. It has eight boundary vertices, whose cycles, beginning with the top vertex, include the partial strings 2; 3,3; 2; 1,3; 4; 1,3; 2; 3,3 respectively. Taking each of these vertices in turn, we test each of the stars containing these strings to see whether the vertex cycle can be completed without violating the proviso or forcing some other vertex to be incompletable. If it passes this test, we then check to see whether this completion is necessarily consistent with the Penrose arrowing. It is easy to check that the arrowing is consistent in every case. For example, a vertex with partial string 1,3 can be completed (only) to a cycle of type 1,2,1,3,3 and we see by inspection that there can be no contradiction (Figure 6.9). Note that a vertex of type $(3,3,...)$ *must* be completed to $(3,3,4)$.

Now let us look at type $(2,2,2,2,2)$. It has ten boundary vertices, with partial strings 3,3, and 2. A vertex with string 3,3 can be completed in two ways, to the cycle $(3,3,4)$ or to $(1,2,1,3,3)$. Let us look at the second of these cycles first. Figure 6.10 shows that after completing this cycle we must next complete a cycle with string 2,4. There are two ways to do this, since both $(2,4,4)$ and $(2,2,2,4)$ contain this partial string. The first choice works; the second does not because it forces a vertex with cycle $(3,3,3,1)$, which is not in the atlas. Discarding this possibility, we are left with $(2,4,4)$, which forces *every* 3,3 vertex of the configuration to be completed to $(1,2,1,3,3)$. This is the arrowing $(2,2,2,2,2)$ (a) of Figure 6.8.

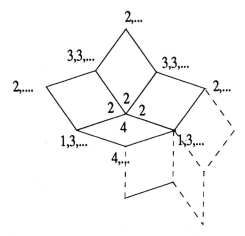

Fig. 6.9 The configuration of vertex type (2,2,2,4) has eight boundary vertices. A vertex with partial string 1,3 must be completed to a cycle of type (1,2,1,3,3), and when this is done the arrowing is consistent.

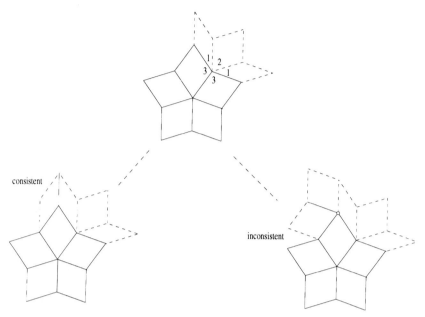

Fig. 6.10 Completing (2,2,2,2,2) by extending the string 3,3, to (1,2,1,3,3).

The case (3,3,4) is only slightly more complicated. This choice forces the cycles of the neighboring vertices to contain the string 1,2. There are three ways to complete a cycle with this string: the full cycle can be (1,2,1,3,3), (1,1,2,2,2,2), or (1,1,2,1,1,2,2). The first choice works with no inconsistency in the arrowing. However, the second cycle gives rise to the forbidden (1,3,3,3) vertex. The third cycle can be realized in two ways, one of which works and one of which must be discarded for reasons of inconsistency. Notice (Figure 6.11) that either of the two admissible configurations forces *every* 3,3 string in this configuration to belong to the cycle (3,3,4); this is arrowing (2,2,2,2,2)(b).

Now let us consider the vertices of a Penrose tiling as a Delone set Λ (Figure 2.7). We will show that the Voronoï cell of any vertex is completely determined by its star.

It is easy to determine the parameters r_0 and R_0 of Λ: the minimum distance between points, $2r_0$, is the short diagonal of a t rhomb, and R_0 is the radius of the largest empty hole in Λ, which is a circle with three vertices of a T rhomb on its boundary. Since the Voronoï cell of a point in a Delone set is contained in a sphere of radius $2R_0$ about the point

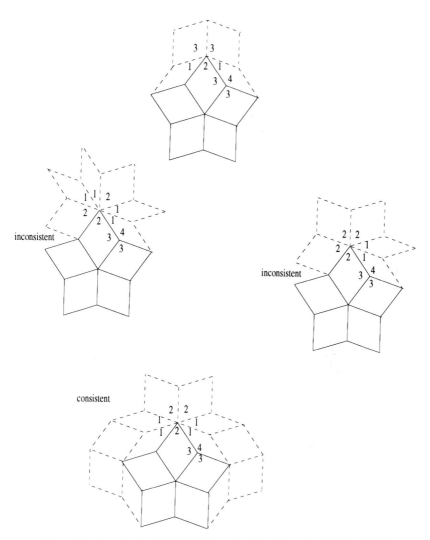

Fig. 6.11 Completing (2,2,2,2,2) by extending the string 3,3, to (3,3,4).

(Corollary 5.2), the Voronoï cell of a Penrose tile vertex is determined by the points within a circle whose radius is less than the length of the long diagonal of a *T* rhomb. By inspecting Figure 6.8 we see that the Voronoï cells are indeed determined by the atlas.

There are seven Voronoï cells, corresponding to the seven (unmarked) vertex configurations (Figure 6.12): three are pentagons, two are

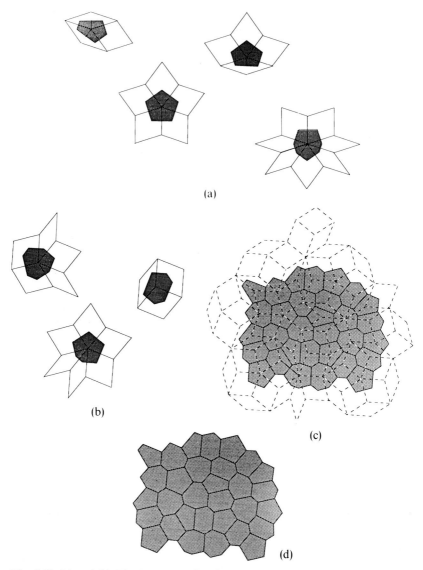

Fig. 6.12 (a) and (b) The Voronoï cells of the vertices of a Penrose tiling. (c) and (d) Part of the Voronoï decomposition of the plane associated to a Penrose tiling.

hexagons, and two are heptagons. There are five edge lengths and three angle measures, $\alpha = 3\theta, \beta = 3.5\theta$, and $\gamma = 4\theta$.

To show that these seven polygons determine the vertex atlas, we need to show that every tiling admitted by them is associated to a Penrose tiling. We can restrict ourselves to edge-to-edge tilings, so that each edge must be matched to another of the same length. Since the sum of the angles at a vertex must be 10θ, the only cycles of angles allowed at any vertex are (α, α, γ) and (α, β, β).

These constraints allow us to establish, by trial and error, that *every* tiling admitted by these prototiles has the Penrose vertex atlas.

Now we turn to other methods of constructing the Penrose tilings. Each of them sheds a somewhat different light on their curious properties. These construction methods were described in detail in Chapter 5; here we show only how they apply to, and help us to understand, the Penrose tilings of the plane.

6.2 Up–down generation

We first show that the Penrose tilings are substitution tilings, from which it follows that the prototiles admit a tiling of the plane.

The composition rule is exhibited in Figure 6.13. Letting T', t' denote the composed rhombs, we see that the rule can be expressed in the form

$$T' = 2T + t, \quad t' = T + t, \tag{6.1}$$

with matrix

$$\mathcal{M} = \begin{pmatrix} 2 & 1 \\ 1 & 1 \end{pmatrix};$$

in fact, $\mathcal{M} = \mathcal{P}^2$, where $\mathcal{P}$ is the Fibonacci matrix (4.10). As we have seen, many of the properties of substitution tilings can be deduced from the composition matrix. For example, this matrix tells us that the ratio of T to t rhombs is $\tau : 1$. You can easily check that the matching rules force the substitution property. By Theorem 5.3, this constitutes a proof that the marked prototiles are an aperiodic set.

Figure 6.13 suggests that the study of the substitution properties of the Penrose tilings would be simpler if we took as our prototiles the triangles of which the rhombs are composed, rather than the rhombs themselves.

To construct the triangles, we divide the T rhomb along its long diagonal and the t rhomb along its short diagonal. This produces two pairs of triangles with mirror image edge markings. We add triple and quadruple arrows to the bisectors to ensure that the triangles will be

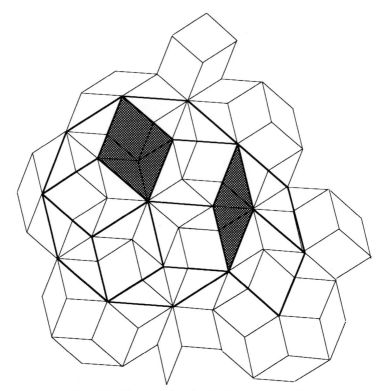

Fig. 6.13 Composition for the Penrose rhombs.

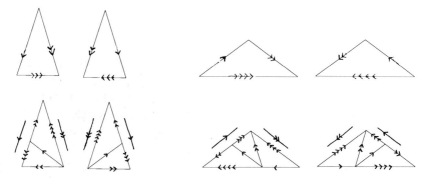

Fig. 6.14 The t_R, t_L, T_L, T_R prototiles and their substitution atlas.

matched properly. We will call the four prototiles *elementary triangles* and denote them by T_L, T_R, t_L, t_R (Figure 6.14); a triangle is an L (R) triangle if its single arrow points in the clockwise (counterclockwise) direction.

Proposition 6.1 Let T'_L, T'_R, t'_L, t'_R be copies of the elementary triangles whose edges have been magnified by the scale factor τ. Each of T'_L, T'_R, t'_L, t'_R can be decomposed into elementary triangles, in exactly one way.

Proof The composition of small triangles into larger ones is shown in Figure 6.15; uniqueness can be established by trial and error. ❏

The larger triangles acquire matching rules from their smaller components as shown in Figure 6.14.

The substitution map for the triangles is thus

$$T'_R = T_R + T_L + t_R, \quad T'_L = T_R + T_L + t_L,$$
$$t'_R = T_L + t_R, \quad t'_L = T_R + t_L;$$

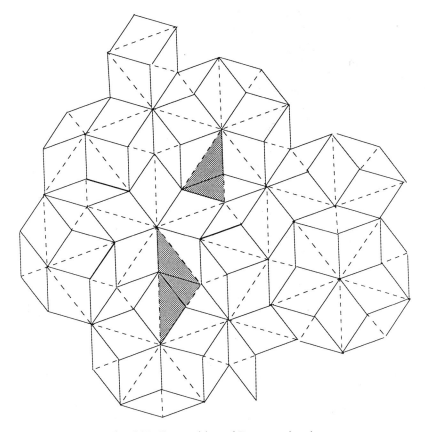

Fig. 6.15 Composition of Penrose triangles.

the matrix is

$$\mathcal{A} = \begin{pmatrix} 1 & 1 & 1 & 0 \\ 1 & 1 & 0 & 1 \\ 0 & 1 & 1 & 0 \\ 1 & 0 & 0 & 1 \end{pmatrix}. \tag{6.2}$$

$\mathcal{A}$ is primitive, since the entries of $\mathcal{A}^2$ are all positive. Its characteristic equation is

$$\lambda^4 - 4\lambda^3 + 5\lambda^2 - 4\lambda + 1 = 0, \tag{6.3}$$

which factors into

$$(\lambda^2 - 3\lambda + 1)(\lambda^2 - \lambda + 1).$$

The eigenvalues are $\tau^2, \tau^{-2}, \exp(\pm i\pi/3)$; the eigenvectors belonging to τ^2 are multiples of $(\tau, \tau, 1, 1)$. Under iteration of $\mathcal{A}$, the eigenvectors belonging to τ^2 elongate, those belonging to τ^{-2} contract, and the eigenvectors belonging to the sixth roots of unity are rotated, in the (real) plane they define, through $\pi/6$. Thus the behavior of the population vector does not have a simple interpretation, although its direction does converge to a multiple of the leading eigenvector.

The composed tiling is again a Penrose tiling. It follows that composition can be iterated *ad infinitum*.

The construction of larger triangles by composition can be regarded as a sequence of embeddings:

$$T_R \xrightarrow{\alpha} T_R \qquad T_R \xrightarrow{\beta} T_L \qquad T_R \xrightarrow{\gamma} t_L$$

$$t_R \xrightarrow{\delta} T_R \qquad t_R \xrightarrow{\epsilon} t_R. \tag{6.4}$$

The mirror image embeddings $\alpha^*, \beta^*, \gamma^*, \delta^*, \epsilon^*$ are obtained by interchanging L and R. We will call these ten embeddings *elementary composition maps*.

The elementary composition maps correspond to the edges of a directed graph (Figure 6.16(a)); any sequence of maps can be represented by a path in the graph. Conversely, every path represents a sequence of embeddings of a given triangle into larger and larger triangles.

An infinite path in the graph is a prescription for constructing a Penrose tiling by 'up–down generation' (Chapter 5). Each embedding is one step 'up'; after contructing it we go 'down' by decomposing the large triangles into elementary ones (Figure 6.16(b)). In this way we associate to any

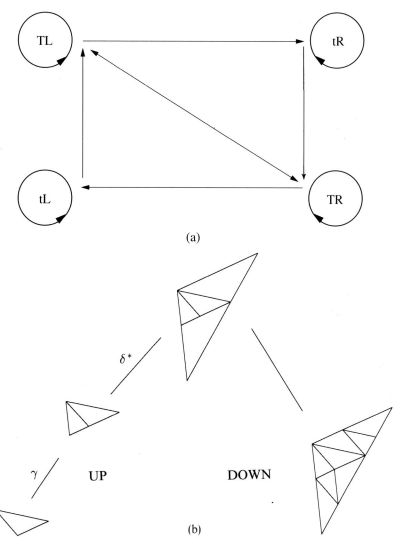

Fig. 6.16 (a) The directed graph. (b) Up – down generation corresponding to the finite path $\gamma \delta^*$.

finite path in the graph a tiling of part of the plane by Penrose triangles (and hence Penrose rhombs). This leads either to a tiling of the entire plane, or to a wedge that can be extended to a tiling of the plane by reflection in its sides (Figure 6.17). Thus we have shown that

Theorem 6.2 The Penrose triangles tile the plane.

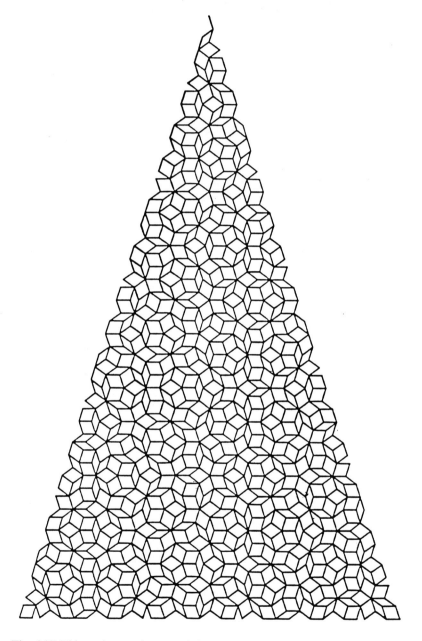

Fig. 6.17 This wedge can be extended to a tiling of the plane by reflection in its sides.

Corollary 6.1 The Penrose triangles are an aperiodic protoset.

Proof This follows directly from the theorem and from Theorem 5.3.

Theorem 6.3 There are uncountably many distinct (i.e. not congruent) Penrose tilings of the plane.

Proof Let us define a *composition sequence* to be an infinite sequence of elementary composition maps $\rho_1, \rho_2, \ldots$ or, equivalently, an infinite path in the directed graph.

By up–down generation, every composition sequence is associated to a Penrose tiling. But every Penrose tiling is associated to many composition sequences since, beginning with any elementary tile, we can determine the unique infinite path in the graph corresponding to successive embeddings.

Thus the correspondence between tilings and composition sequences is not one to one. However, the relation between them is not hard to determine. Let $\rho = \{\rho_0, \rho_1, \rho_2, \ldots\}$ be a composition sequence, and let C be the composition operator or shift map:

$$C(\{\rho_0, \rho_1, \rho_2, \ldots\}) = \{\rho_1, \rho_2, \rho_3, \ldots\}. \tag{6.5}$$

We will say that two composition sequences are *cofinal* if they agree after a finite number of terms.

Lemma 6.1 Two composition sequences correspond to the same tiling if and only if they are cofinal.

Proof Let ρ and ρ' be two composition sequences, and T and T' the tilings they define. If $\rho = \rho'$ then $T = T'$. If ρ and ρ' are cofinal, then they describe the same tiling beginning at some hierarchical level; let us call this tiling T^*. But since composition is unique, T^* can be decomposed into elementary tiles in only one way, so in fact $T = T'$. Conversely, suppose ρ and ρ' define the same tiling T. Let the elementary maps with which they begin be ρ_0 and ρ_0'. These elementary maps have as initial tiles T_0 and T_0', both of which belong to T. Since the distance between T_0 and T_0' must be finite, at some hierarchical level there will be a single tile that contains both of them. From that level on, the two composition sequences must be the same. ❏

Cofinality is an equivalence relation on the set of composition sequences. We will call each equivalence class a *family*. Obviously, composition preserves families:

Lemma 6.2 Two sequences ρ and ρ' are cofinal if and only if $C(\rho)$ and $C(\rho')$ are also cofinal.

Notice that, in general, $C(\rho) \neq \rho$; thus, in general, Penrose tilings are not 'self-similar'. However, since a composed tiling is again a Penrose tiling, composition maps the set of all Penrose tilings to itself.

Using the directed graph, we can conclude the proof of Theorem 6.3 by invoking the standard Cantor diagonalization argument: if there were only countably many families of sequences then, using the Axiom of Choice, we could select one member of each family and create a countable list of family representatives. But this list would necessarily be incomplete since – with the aid of the directed graph – we could always construct a sequence that is not on it. Thus the number of Penrose tilings is uncountable. ❑

6.3 Pentagrids and their duals

For a very different view of the Penrose tilings, we study their duals in the form of de Bruijn's *pentagrids*. Pentagrids are the special cases $n = 2, k = 5$ of the multigrid construction discussed in Chapter 5.

Thus a pentagrid is a superposition of five grids, or infinite sets of equally spaced parallel lines, where each grid is orthogonal to one of the vectors of the grid star which, in the case of the pentagrids, point to the vertices of a regular pentagon. In order for the pentagrid to be dual to a tiling of the plane by quadrilaterals only, we must shift the jth grid, $j = 1, \ldots, 5$ through some distance γ_j in the direction of the grid vector $\vec{e}_j$ so that at most two lines of the pentagrid intersect in any point. A pentagrid so shifted is said to be *regular*. A portion of a regular pentagrid is shown in Figure 5.19.

For a fixed grid star, the pentagrids are completely characterized by the quintuple of real numbers $(\gamma_1, \ldots, \gamma_5)$. Two quintuples $(\gamma_1, \ldots, \gamma_5)$ and $(\gamma'_1, \ldots, \gamma'_5)$ define the same pentagrid if and only if $\gamma_j - \gamma'_j \in Z$, $j = 1, \ldots, 5$. (Certain choices of the γ_js correspond to translations of the grid, others to nonsuperposable grids; see Section 6.4.3.)

A key idea of the pentagrid analysis of the Penrose tilings is to focus on the regular case, which is generic, treating the singular cases as limits of regular ones. This is in contrast to the usual discussion of kite and dart tilings (see e.g. G&S), which focuses on special symmetrical cases.

As explained in Chapter 5, a regular pentagrid is a blueprint for its dual rhomb tiling. The vectors $\vec{e}_1, \vec{e}_2, \ldots, \vec{e}_5$ of the grid star define two rhombic prototiles, up to rotation: those with edges parallel to $\vec{e}_j$ and $\vec{e}_{j+1}$, and those with edges parallel to $\vec{e}_j$ and $\vec{e}_{j+2}$ (subscripts are reduced modulo 5). We associate to every intersection point of the pentagrid the rhomb whose edges are the two star vectors orthogonal to the intersecting lines.

The dual tiling is thus a tiling by T and t rhombs, but it does not necessarily satisfy Penrose's rules. De Bruijn showed that one additional condition on $(\gamma_1, \ldots, \gamma_5)$ is necessary (and sufficient) to produce a Penrose tiling; let us see what it is and why it works. We saw in Chapter 5 that every mesh, and thus every vertex in the dual tiling, can be labeled with a k-tuple of integers. When $k = 5$, the inequalities (5.6) become

$$-\frac{1}{2} < l_j + \gamma_j - \vec{x} \cdot \vec{e}_j < \frac{1}{2}, \quad j = 1, \ldots, 5. \tag{6.6}$$

Adding these five inequalities, we obtain the single condition

$$0 < \sum_{j=1}^{5} l_j + \sum_{j=1}^{5} \gamma_j + \frac{5}{2} - \vec{x} \cdot \left(\sum_{k=1}^{5} \vec{e}_j \right) < 5. \tag{6.7}$$

Since the vectors $\vec{e}_j$ point to the vertices of a regular polygon, their sum is zero. Thus (6.7) reduces to

$$0 < \sum_{j=1}^{5} l_j + \sum_{j=1}^{5} \gamma_j + \frac{5}{2} < 5. \tag{6.8}$$

Now (6.8) is satisfied by every pentagrid and pentagrid dual. But in the special case when

$$\sum_{j=1}^{5} \gamma_j = -5/2, \tag{6.9}$$

then we can simplify further: (6.8) reduces to

$$0 < \sum_{j=1}^{5} l_j < 5.$$

Since the l_j are integers, we have

$$\sum_{j=1}^{5} l_j \in \{1, 2, 3, 4\}. \qquad (6.10)$$

We will call this sum the *index* of the vertex that corresponds to the mesh. (The indices are the numbers attached to the vertices in Figure 6.6.) We have shown that

Proposition 6.2 Every vertex in a pentagrid dual tiling (with $\sum_{j=1}^{5} \gamma_j = -5/2$) has index 1, 2, 3 or 4.

Since adding an integer to any γ_j does not change the pentagrid or its dual, we can replace $-5/2$ by $1/2$ in (6.9).

Definition 6.1 The condition $\sum_{j=1}^{5} \gamma_j = 1/2$ (modulo 1) is called the *sum condition*.

Now we will prove that the sum condition implies that the dual of a regular pentagrid is a Penrose tiling.

Lemma 6.3 If a pentagrid satisfies the sum condition, then the indices of the four vertices of a rhomb in the dual tiling can only be, in cyclic order, $(1, 2, 3, 2)$ or $(2, 3, 4, 3)$.

Proof As we pass from one mesh of the pentagrid to another by crossing a common edge, we pass from the nth strip of the jth grid to $(n+1)$st or $(n-1)$st strip of the same grid, so only l_j is changed, increasing or decreasing by 1. Thus as we follow a small circle around an intersection point of the jth and kth grids, the kth and jth components of the location vector first increase by one and then decrease by the same amount, returning us to the starting vertex and starting index (see Figure 6.6). ❑

De Bruijn proved that the tilings that satisfy the sum condition are true Penrose tilings by showing that they can be equipped with Penrose matching rules. Let us assume that we are given the dual of a regular pentagrid that satisfies the sum condition, with indices attached to the vertices. There are three classes of edges in a pentagrid dual: those joining vertices of index 1 to vertices of index 2, those joining vertices of index 2 to vertices of index 3, and those joining vertices of index 3 to vertices of index 4. We will show how the indices provide the information needed to decorate the edges with arrows. First, we place single arrows on the edges joining indices 1 and 2, all pointing from 2 to 1. This choice dictates that the edges joining indices 3 and 4 must be decorated with single arrows pointing from 3 to 4. To show that we have a Penrose tiling, we want to put double arrows on remaining edges, those joining indices 2 and 3. But can this be done consistently? Since the single-arrow edges have already been determined, the directions of the double arrows *on each rhomb* are also determined. The problem is to show that *in the tiling*, the arrows on any two rhombs that share a double-arrowed edge must point in the same direction.

This is the same sort of question we confronted in the proof of Theorem 6.1. Let us first ask under what conditions the arrows might point in opposite directions. Figure 6.18 shows the three possiblities. In fact none of them can arise in a pentagrid dual.

The reason, which is built into the pentagrid construction (see de Bruijn, 1987a; Socolar, 1990b), is that these duals must satisfy an 'alternating condition': *mirror-image rhombs must alternate in every ribbon*. The configurations in Figure 6.18 (a) and (c) obviously fail to satisfy this condition; with a little more work you will see that the third configuration (b) can be extended only to a configuration in which the condition also fails.

The reason why the alternating condition must hold is simple; a general proof can be based on the following observation. The vectors $\vec{\epsilon}_j, j = 1, 2$ are the mirror images of the vectors $\vec{\epsilon}_{5-j}$ with respect to the line containing $\vec{\epsilon}_3$. Let ℓ be any line of grid 3 and consider the pattern of intersections of the lines of grids 1 and 5 with ℓ (Figure 6.19). The intersections of ℓ with the lines of grid 1 correspond, in the dual, to t rhombs which are the mirror images of the rhombs corresponding to the intersection of the lines of grid 5 with ℓ; similarly, the intersections of the lines of grids 2 and 4 correspond to mirror image T rhombs.

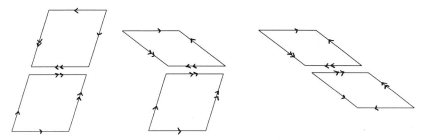

Fig. 6.18 Double arrows pointing in opposite directions.

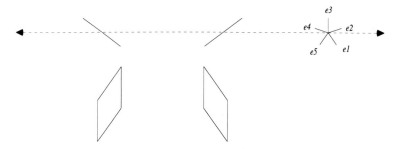

Fig. 6.19 The intersections of the lines of grids 1 and 5 with ℓ correspond to mirror image tiles in the dual tiling.

Each family of lines cuts ℓ in a sequence of equispaced points; the spacings for grids j and $5 - j$ are the same. It follows that the intersections of the lines of the jth and $(5 - j)$th grids with ℓ must alternate with one another, and thus in any ribbon of the tiling mirror image rhombs must also alternate. Obviously, we can replace $\vec{\epsilon}_3$ with any of the other star vectors. Thus we conclude that

Theorem 6.4 If the shift vector $\vec{\gamma}$ of a pentagrid is regular and satisfies the sum condition then the dual tiling is a Penrose tiling.

In his remarkable paper, de Bruijn also showed that *every* Penrose tiling is either the dual of a regular pentagrid (that satisfies the sum condition), or else it can be uniquely associated to the dual of a singular pentagrid.

The pentagrid construction reveals the reason why locally legal configurations may not be globally legal. The pentagrid is a blueprint for the Penrose tiling: it completely determines the relative positions of the tiles. Each line of each grid intersects *all* of the lines of the other grids. When we create a patch of tiles, we are specifying the positions of several lines of several grids and thus implicitly the positions of possibly distant intersection points. At the same time, we are also specifying the positions of tiles that are *not necessarily adjacent to the original tile configuration*.

This is the problem of 'empire', discussed in detail (for kite and dart tilings) in G&S. In fact, although the matching rules appear to give the tiler a large number of choices at every stage of the construction, the Penrose tilings are almost completely determined by a small number of them. In Section 6.7 we will show that a Penrose tiling is almost completely determined by only two of its vertices! (Of course, not *any* two vertices: they must be properly chosen.)

6.4 Penrose tilings as projections

As we saw in Chapter 5, the multigrid construction has a 'natural' interpretation in a higher-dimensional space. In the case of the pentagrids, that higher dimension is five.

6.4.1 The canonical projection

It follows from Hadwiger's theorem (Theorem 2.1) that the pentagrid star (of vectors $\vec{\varepsilon}_1, \ldots, \vec{\varepsilon}_5$) is an orthogonal projection, onto a two-dimensional plane, of the vectors of an orthonormal basis $\vec{e}_1, \ldots, \vec{e}_5$ of E^5:

$$\vec{e}_j \longrightarrow \vec{\varepsilon}_j, \; j = 1, \ldots, 5. \tag{6.11}$$

When we reinterpret the pentagrid construction in five-dimensional space, the shift vector $\vec{\gamma} = (\gamma_1, \ldots, \gamma_5)$ becomes a translation in E^5, the location vector $\vec{k} = (k_1, \ldots, k_5)$ becomes a point of the integer lattice $\mathcal{I}_5$, and the grid equations (5.5) now represent equispaced families of parallel hyperplanes orthogonal to the basis vectors $\vec{e}_j$. The five sets of grid hyperplanes partition E^5 into five-dimensional hypercubes; each hypercube is the Voronoï cell of the lattice point at its center. The plane pentagrid we studied in the previous section is just the intersection of a plane $\mathcal{E}$ with the hypercube tessellation! The requirement that $\vec{\gamma}$ must be regular means that $\mathcal{E}$ does not meet any vertex, edge, or 2-face of the (translated) Voronoï tessellation.

We know that $\vec{k} \in \mathcal{I}_5$ is the location vector of a mesh in the pentagrid if and only if $\mathcal{E}$ intersects the interior of the translate, by $\vec{\gamma}$, of its Voronoï cell $V(\vec{k})$. By Proposition 2.17, this is equivalent to requiring

$$\Pi^\perp(\vec{k}) \in \Pi^\perp(V(0) + \vec{\gamma}). \tag{6.12}$$

This reformulation allows us to construct the Penrose tilings by canonical projection.

As we saw in Chapter 2, $\mathcal{I}_5$ is invariant under rotation through 2θ about $\vec{w} = (1, 1, 1, 1, 1)$, the body diagonal of the unit hypercube (recall that $\theta = \pi/5$). This rotation, which permutes $\vec{e}_1, \ldots, \vec{e}_5$ cyclically, has two invariant totally irrational planes orthogonal to one another and to the fixed 'axis' $< \vec{w} >$, the one-dimensional subspace generated by $\vec{w}$.

We choose for $\mathcal{E}$ the plane that is rotated through an angle of 2θ (the other, $\mathcal{E}'$, is rotated through 4θ). The orthogonal complement of $\mathcal{E}$, $\mathcal{E}^\perp$, is three-dimensional: it is the direct sum of $\mathcal{E}'$ and $< \vec{w} >$.

The projection principle was explained in Chapter 5: we project, onto $\mathcal{E}$, the duals of the facets of the Voronoï tessellation $\mathcal{V}(\mathcal{I}_5^p) + \vec{\gamma}$ that are cut by $\mathcal{E}$. Since $\mathcal{E}$ meets only faces of $\mathcal{V}(\mathcal{I}_5^p) + \vec{\gamma}$ of dimensions 3, 4 and 5, the corresonding faces of the Delone tiling have dimensions 2,1, and 0; in fact, the 2-faces form a connected surface of square faces. The projections of these squares onto $\mathcal{E}$ are the rhombs of the Penrose tiling.

The set X of lattice points that we project onto $\mathcal{E}$ (where they become the vertices of the rhombs) are precisely those that project, under $\Pi^\perp$, into the window $K = \Pi^\perp(V(0) + \vec{\gamma})$.

6.4.2 *The vertex atlas*

Under the duality discussed above, the star of a vertex v of a Penrose tiling corresponds to the intersection of $\mathcal{E}$ with a cluster of Voronoï cells in E^5. We can describe the cluster precisely. Suppose that $v = \Pi(x)$, $x \in \mathcal{I}_5^p$. By (6.12), we know that $\Pi^\perp(x) \in \Pi^\perp(V(0) + \vec{\gamma})$. In particular, $\Pi^\perp(x)$ must lie in the intersection of several projected facets of $V(0) + \vec{\gamma}$. Let $\mathbf{f}_1, \ldots, \mathbf{f}_j$ be these facets and $\vec{f}_1, \ldots, \vec{f}_j$ the corresponding facet vectors (see Chapter 2). Then since $V(x)$ tiles E^5 by translation, $\mathbf{f}_1 - \vec{f}_1, \ldots, \mathbf{f}_j - \vec{f}_j$ are also facets of $V(0) + \vec{\gamma}$, so $\Pi^\perp(x - \vec{f}_1), \ldots, \Pi^\perp(x - \vec{f}_j) \in \Pi^\perp(V(0) + \vec{\gamma})$. This means that $\mathcal{E}$ cuts $V(x - \vec{f}_1), \ldots, V(x - \vec{f}_j)$.

Thus decomposition of the window K by the projected facets of $V(0)$ determines the vertex configurations in the projected tiling, and also their relative frequencies (see Section 4.3). The window K is a rhombic icosahedron; it is the projection of the unit hypercube $\mathcal{Q}_5$ into the three-dimensional subspace generated by $\mathcal{E}'$ and $< \vec{w} >$. The window is not densely filled with projected lattice points: $\vec{w}$ is a lattice vector and so, for every $\vec{x} \in \mathcal{I}_5$ we have $\vec{x} \cdot \vec{w} \in Z$. In particular, the 22 vertices of K, which are a subset of the projected vertices of $\mathcal{Q}_5$, satisfy $\vec{v} \cdot \vec{w} \in \{0, 1, 2, 3, 4, 5\}$. The points of X are projected into the four cross-sections of K orthogonal to $\vec{w}$ at levels $\{1, 2, 3, 4\}$. Recall that these levels are precisely the values of the indices of the rhombs in the pentagrid construction; the index of a vertex is just the level on which the preimage of the vertex lies in E^5. The levels are shown in Plate 1 (top) facing p. 196.

When we project $\mathcal{Q}_5$ into $\mathcal{E}^\perp$, its ten facets are projected into K. The facets are four-dimensional cubes; their projections are rhombic dodecahedra. The top left polyhedron in Plate 2 facing p. 197 is K; on the right we see one projected facet. The projected facets overlap in K; two of them, related through the center of K, are shown at the bottom left in this diagram. The tiny polyhedron in the center is the intersection of the ten dodecahedra. We see this polyhedron more clearly on the right; in this

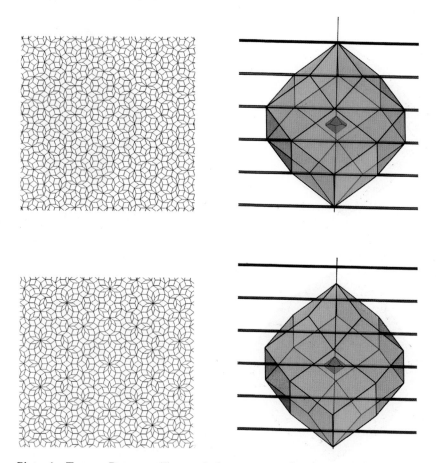

Plate 1 Top: a Penrose tiling and the corresponding cross-sections of K.
Bottom: a generalized Penrose tiling obtained by setting $\varpi = 0$.

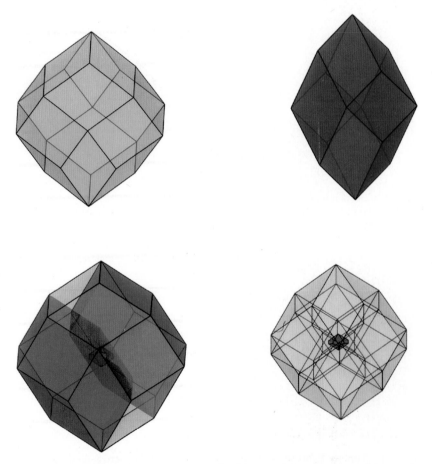

Plate 2 Top left: the window $K = \Pi^{\perp} V(0)$; top right: a projected facet of $V(0)$; bottom left: two of the facets in K, together with the polyhedron that is the intersection of all ten facets; bottom right: here all ten facets are shown (as edge graphs), together with their intersection.

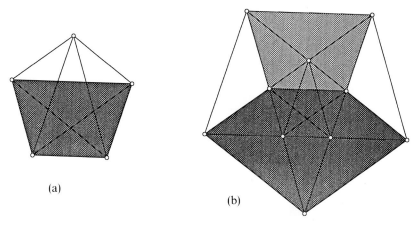

(a)

(b)

Fig. 6.20 The Penrose cross-sections of K. (a) Level 1. The shaded region is a section of one of the five dodecahedra that share the vertex at level 0. (b) Level 2. The darker region is the section of the same dodecahedron as in (a); the lighter region is the section of a dodecahedron with a vertex at level 5. (These are the blue and red dodecahedra of Plate 2.)

diagram the edges (but not the faces) of all ten dodecahedra are also shown.

Since the points $\Pi^{\perp}(X)$ lie in four planar cross-sections of K, these sections are the only features of K that we need to study. The two types of cross-section that we obtain in the Penrose case are shown in Figure 6.20. Both are pentagons. The pentagon in (a) is the cross-section at level 1 (or level 4, which is the image of level 1 through the center of K); the pentagon in (b) is the cross-section at level 2 (or 3). The heavy lines indicate the section of a single dodecahedron.

At level 1, the 11 regions into which the pentagon is decomposed – a regular pentagon and ten triangles, similar to the T and t triangles – are cross-sections of the five dodecahedra which share the vertex at level 0. The pentagon in the center corresponds to the intersection of the cross-section plane with all five dodecahedra and thus to the vertex configuration $(2,2,2,2,2)$. The t triangles correspond to the intersection of four dodecahedra and thus to vertex configurations of the type $(2,2,2,4)$. The T triangles are the intersection of the cross-section plane with three dodecahedra; there are two possible configurations, with vertex cycles $(4,4,2)$ and $(3,3,4)$. But if we remember how these configurations must be arrowed (Figure 6.8) and how the arrows are related to the levels (Figure 6.6), we see that the T triangles correspond to the configuration $(3,3,4)$.

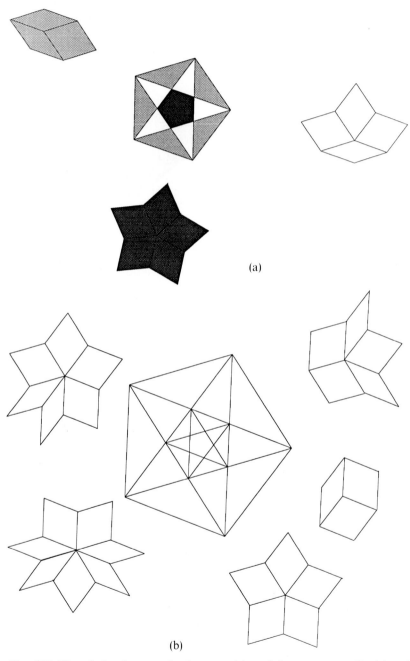

Fig. 6.21 The relation between the decomposition of the pentagon at level 1 and the Penrose vertex atlas is shown explicitly in (a); in (b), matching the regions at level 2 with the remaining vertex configurations is left as an exercise.

At level 2, all ten dodecahedra come into play (though there is no common overlap region). Through considerations like those above, we can deduce the correspondences shown in Figure 6.21(a). (Test your understanding by finding the correspondences in (b).)

6.4.3 The role of the shift vector

Changing the shift vector translates K and thereby changes X: let's see what happens to it. Assume that some initial $\vec{\gamma}$ is given. It is helpful to think of $\vec{\gamma}$ in terms of the orthonormal basis $\vec{u}_1, \ldots, \vec{u}_5$ for E^5 that we constructed in Chapter 2. Recall that $\vec{u}_1, \vec{u}_2$ span $\mathcal{E}$, $\vec{u}_3, \vec{u}_4$ span $\mathcal{E}'$, and $\vec{u}_5$ is parallel to $\vec{w}$. Thus $\vec{\gamma}$ is a linear combination of three vectors:

$$\vec{\gamma} = \alpha\vec{v} + \beta\vec{v}' + \varpi\vec{w}, \tag{6.13}$$

where $\vec{v} \in \mathcal{E}$ and $\vec{v}' \in \mathcal{E}'$.

The sum condition says that we must have $\varpi = 1/2$ (modulo 1) in order for the corresponding tiling to be a Penrose tiling $\mathcal{T}$. Changing the $\mathcal{E}$ component of $\vec{\gamma}$ simply translates $\mathcal{T}$, since the set of lattice points that we project is left unchanged. However, when the $\mathcal{E}'$ component is changed some points of $\mathcal{I}_5^p$ are removed from the cylinder and others enter into it (as in Figure 2.11). Thus we obtain a different Penrose tiling. Figure 6.22 shows how the Penrose tiling can change – locally – with $\vec{\gamma}$.

If we change ϖ by a non-integer quantity, then the sum condition is no longer satisfied and the projected tiling is no longer a Penrose tiling. It is (hopefully) clear from Plate 2 why vertex configurations prohibited in the Penrose case will now appear. Tilings for which $\varpi \neq 1/2$ are called *generalized* Penrose tilings. Figure 6.23 shows the central cross-section corresponding to Plate 1 (bottom), where $\varpi = 0$. This slice cuts through the center of the overlap region of the ten dodecahedra (Figure 6.23) and thus the configuration of ten t rhombs will appear in the tiling. Their frequency is proportional to the area of the cross section of the core, which is maximized when $\varpi = 0$.

The Penrose case and the case $\varpi = 0$ are the only two in which pairs of cross-sections are images through the center of K. Thus almost all generalized Penrose tilings are considerably more complicated than these. But, since they are generated by canonical projection, all generalized Penrose tilings are crystals (see Chapter 8).

It is not known which, if any, of the generalized Penrose tilings are substitution tilings, but it is known that matching rules exist for a countable family of them. Very simple matching rules for the $\varpi = 0$ case

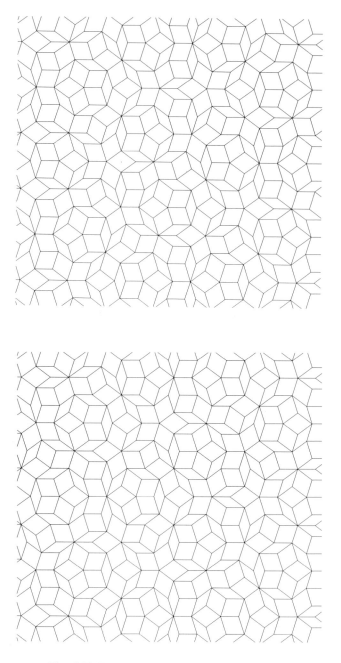

Fig. 6.22 Two Penrose tilings with different $\vec{\gamma}$s.

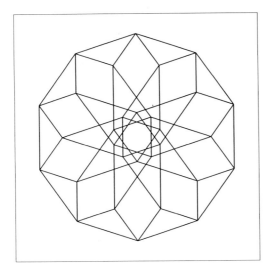

Fig. 6.23 The central cross section of K when $\varpi = 0$.

(see Figure 7.15) have been found recently by Le (1993); they will be discussed in Chapter 7.

Using the interactive computer program 'QuasiTiler', you can discover the properties of the tilings associated to various choices of $\vec{\gamma}$ (see Section 8.4.4).

6.5 Other generalizations of the Penrose tilings

In addition to disobeying the sum condition, as in the preceding section, we can generalize the Penrose tilings in other ways. In some cases we obtain tilings that are not crystals. We describe one such tiling family – the 'binary' substitution tilings – briefly here.

The binary substitution tilings are a two-dimensional analogue of the LB sequences discussed in Chapter 4. The prototiles of the version we will consider are again the T and t rhombs, but the substitution matrix is $\begin{pmatrix} 3 & 1 \\ 1 & 2 \end{pmatrix}$, the matrix for the LB sequences. It follows that the relative frequencies of the T and t rhombs are *exactly the same as in the Penrose case*, but the tilings are very different indeed. In fact, the binary tilings are not crystals. Notice (Figure 6.24) that the tiles of this tiling do not obey the alternation rule; thus the tiling is not the dual of a pentagrid. However, the edges of the tiles are the arms of the star of Figure 6.19, and so the

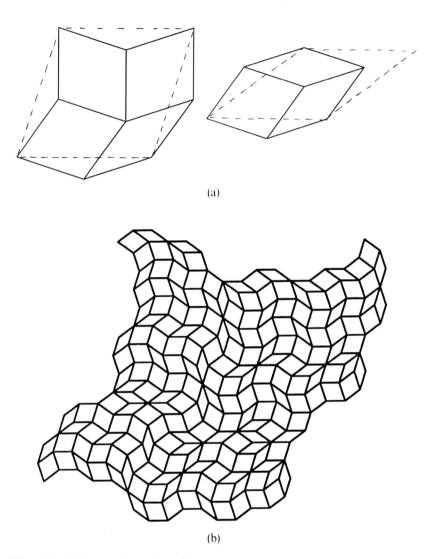

(a)

(b)

Fig. 6.24 A binary substitution tiling. (a) Decomposition of the prototiles; (b) a portion of a binary tiling.

tiling can be lifted to a surface of squares with vertices on the point lattice $\mathcal{I}_5^p$. Since the leading eigenvalue of the substitution matrix is not a Pisot number, this surface will not stay close to the plane $\mathcal{E}$. A diffraction pattern produced by this tiling is shown in Chapter 8.

6.6 The space $\mathcal{P}$

The set $\mathcal{P}$ of all Penrose tilings by rhombs is a 'mathematical object' in its own right. It reveals many things about Penrose tilings that are not so easy to understand by looking at individual tilings. For example, the meaning of 'self-similarity' for the Penrose tilings was clarified when we looked at the family of all Penrose tilings. One can also study $\mathcal{P}$ from the point of view of dynamical systems (see Chapter 7). In this section we identify $\mathcal{P}$ with the space of shift vectors, and outline a proof of the remarkable fact that

Theorem 6.5 (de Bruijn) A Penrose tiling is almost completely determined by two of its vertices.

Proof (sketch) To establish this theorem we need a measure on $\mathcal{P}$ of the 'closeness' of tilings. This makes sense only if we fix a common origin for all the tilings: otherwise we would have to conclude that all Penrose tilings are virtually identical, since they all contain identical patches of arbitrarily large size. Several measures have been proposed (see e.g. de Bruijn, 1992; Radin and Wolff, 1992; Robinson, 1993) and it is not obvious that they are equivalent. For example, one can measure the 'closeness' of two tilings as the size of neighborhood of the origin on which they agree, or as the measure of the set on which they disagree. De Bruijn used the second of these in his proof of this theorem.

Definition 6.2 (de Bruijn) Let Λ be a Delone set, let $\Lambda_1 \subset \Lambda$, and let $\epsilon > 0$. The subset Λ_1 is ϵ-complete in Λ if there exists $q > 0$ such that, for every $x \in E^n$ and $r > q$, we have

$$\frac{|\Lambda_1 \cap B_x(r)|}{|\Lambda \cap B_x(r)|} > 1 - \epsilon.$$

Definition 6.3 The Delone sets Λ and Λ' are equal up to ϵ if $\Lambda \cap \Lambda'$ is ϵ-complete in $\Lambda \cup \Lambda'$.

Before proving the theorem, we need to show that

Lemma 6.4 If two shift vectors $\vec{\gamma}$, $\vec{\gamma}'$ are sufficiently close in E^n then the corresponding projected sets are equal up to ϵ.

Proof (sketch) Let T and T' be two Penrose tilings and $\vec{\gamma}$ and $\vec{\gamma}'$ their shift vectors. We may assume that they differ only in their $\mathcal{E}'$ components. The main idea is that since the pentagonal cross-sections at each of the levels 1,2,3,4 of the two windows K and K' have large overlap if $|\vec{\gamma} - \vec{\gamma}'|$ is small, the area of the overlap is a measure of the overlap of the corresponding cylinders and thus of the projected point sets.

Continuing with the proof of the theorem, we choose a Penrose tiling and two points $x, y \in X$ where, as usual, X is the set of lattice points that we will project. We ask to which translates of the window $\Pi^{\perp}(x)$ and $\Pi^{\perp}(y)$ could both belong. The closer $\Pi^{\perp}(x)$ and $\Pi^{\perp}(y)$ are to one another, the larger will be the set of windows that contain them both. Conversely, the farther they are apart, the smaller this set of windows will be. Let κ be diameter of the window (the least upper bound of the distances between points of $\Pi^{\perp}(X)$). By choosing points x and y in the cylinder with $|\Pi^{\perp}(x) - \Pi^{\perp}(y)|$ sufficiently close to κ, we can ensure that the shift vectors of two windows both containing $\Pi^{\perp}(x)$ and $\Pi^{\perp}(y)$ will differ by an arbitrarily small amount. The theorem then follows from Lemma 6.4. ❑

In fact, the proof is quite general; the theorem is true for *any projected set*, as long as one can reasonably define the least upperbound κ of the distances between projected points in the window K. This can always be done if K is convex, which is the case for the canonical projections and most of the others that have been studied.

6.7 Notes

1. We have not discussed 'Ammann bars', the line configurations one obtains when the Penrose tiles are marked as shown in Figure 6.5 (c), partly because they are discussed in detail in G&S (for kite and dart tilings) and partly because they are closely related to the pentagrids (Socolar and Steinhardt, 1986; de Bruijn, 1994).

2. A reader familiar with the theory of substitution dynamical systems can easily translate all of Section 2 into that language. For an explicit discussion of the Penrose tilings as a dynamical system, see (Robinson, 1993).

3. De Bruijn's paper (1981a) was based on the properties of the algebraic number field $Z[\sqrt{5}]$, a point of view that is receiving increasing attention. See, for example, Coxeter (1993) and Berman and Moody (1994).

4. Physicists use tiling models to study the stability of quasicrystals, in addition to other properties such as their diffraction spectra and their geometry. We briefly describe the context in which the binary tilings were discovered.

'Motivated by the desire to see whether an equilibrium quasicrystalline phase can exist in a suitably chosen simple atomic model with two-body central forces, they (Widom, Strandburg, and Swendson, 1987) constructed a two-dimensional model which encourages local decagonal order', explains Katherine Strandburg (1989). 'When cooled from high temperature using a Monte Carlo technique, the model system solidified into a quasicrystalline phase possessing long-range bond-orientational order and well-ordered nonperiodic atomic planes.' This 'low-temperature' pattern can be interpreted as a tiling of the plane by thick and thin rhombs congruent to the Penrose rhombs.

Although these tiles did not obey Penrose's matching rules, it turned out that they did satisfy a matching condition that looks very much like Penrose's colored vertex rules for kites and darts. Recall that each of the two rhombs has two vertices that are even, and two that are odd, multiples of $\theta = \pi/5$. Assign one color to the odd vertices and another to the even ones. The matching rule is simply that colored vertices must be matched. Tilings that satisfy these rules are called *binary tilings*. The binary rule is not strong enough to force nonperiodicity.

Subsequently it was discovered that a subclass of the binary tilings has the substitution property discussed above.

A second interesting subclass of binary tilings has been studied by Burkov (1992), who argues that instead of joining the atoms to form rhombs the atoms should be linked to form congruent clusters. Thus instead of interpreting the pattern of points as vertices of a tiling of the plane, we should think of them as vertices of a *covering*. More abstractly, the clusters can be regarded as decagons. Two decagons that are not disjoint may share an edge or overlap. The former case can be associated to a t rhomb, the latter to a T rhomb.

The decagonal clusters define a subclass of the binary tilings. Connecting the centers of overlapping clusters, we obtain a graph whose edges have two lengths, l and s. The Voronoï decomposition of the set of vertices of this graph is a tiling of the plane by triangles, trapezoids, and pentagons; this tiling is dual to a binary tiling. (But not all binary tilings are duals of cluster graphs, since the binary rules permit vertex valences other than 3, 4, and 5.)

The cluster model is useful for studying energy properties of real quasicrystals, a subject far beyond the scope of this book. We will just note that it appears to be a useful model. In fact, Burkov has found that 'the free energy as a function of the alloy composition has a cusp at a point exactly corresponding to (this model of) the decagonal quasicrystal'; he argues that this helps to explain the existence of the quasicrystalline phase. Diffraction patterns produced by this model are consistent with experimental data.

Although packings, coverings, and tilings are closely related mathematical subjects, covering models are rarely used in mathematical crystallography. This is probably because coverings appear to contradict the standard 'hardcore' assumption – that atoms do not overlap. In Burkov's model, however, it is the atomic configurations, not individual atoms, that overlap, and molecular structures with interlocking rings are hardly a novelty in crystal chemistry! The cluster model suggests that the geometry of coverings should become part of the mathematical toolbag for crystallography.

5. Penrose tiles can be obtained from Pentaplex Ltd., Royd House, Birds Royd Lane, Brighouse, West Yorks HD6 1LQ (Great Britain).

7

The aperiodic zoo

The mathematical universe is inhabited not only by important species but also by interesting individuals.   *(C. L. Siegel)*

To pause at this point and ask whether one ought to say that there is really one world or five is reasonable enough.Our own view is that the most likely account reveals that there is a single, divine world; different considerations might lead to a different view, but they may be dismissed. Plato   *(Timaeus)*

In this chapter we describe some of the residents of the little zoo of aperiodic tilings, and something of what is known about the world to which they belong. While some of these tilings are currently being studied as models for aperiodic crystals, possible applications will not be our chief concern. Instead, we will be interested in what aperiodic tilings may have to teach us about long-range order. Since, as of this writing, much of the material discussed in this chapter is still unpublished and many fundamental ideas are still being clarified, this chapter – even more than the others – should be regarded as an account of work in progress.

At its present stage of development, the theory of aperiodic tilings is a mix of the general and the particular. Several famous tilings – the Penrose tilings, the Ammann tilings, and a few others – continue to serve as examples of the properties all aperiodic tilings might display, and as a guide for generalizations.

7.1 Matching rules and their classification

A matching rule, or set of matching rules, for a set of tiles is essentially a list of instructions for putting the tiles together (see Definition 5.14). We have seen that matching rules can be realized in many different ways: as a set of instructions accompanying each tile, as an atlas of allowed configurations, and so forth. In particular, as Radin persuasively points out (private communication),

Proposition 7.1 Transitive matching rules can always be realized through the geometry of the prototiles by suitably deforming their facets.

(A proof can be based on the argument in Radin, 1994.)

Of course, in order to express the rules in this way the number of shapes may have to be increased considerably. For example, Penrose's first set (Figure 6.1) of aperiodic prototiles can be realized as four shapes with the regular pentagons marked in three different ways, or as six shapes with no markings. There is no 'best' choice; different realizations of the rules are useful in different contexts.

From now on, by 'matching rules' we will mean rules that force non-periodicity. There are at least two different ways to classify them. The first measures how 'strong' they are. Assume we are given a set P of prototiles that obey a list L of rules for matching the tiles along their facets (the matching need not be facet-to-facet). Borrowing terminology – but not the corresponding definitions – from the physics literature, we might define

Definition 7.1   A set L of matching rules for a protoset P is perfect if all of the tilings admitted by P are nonperiodic and repetitive and belong to a single local isomorphism class. The set L is strong if all of the tilings admitted by P are nonperiodic and repetitive, but do not necessarily belong to the same local isomorphism class. The set L is weak if all of the tilings admitted by P are nonperiodic, but not necessarily repetitive.

In the following section we will describe families of tilings by a curious three-dimensional aperiodic monotile whose matching rule – for some values of its parameter – is weak but not strong. The Klèman–Pavlovitch rules for generalized Penrose tilings are conjectured to be strong but not perfect (Klèman and Pavlovitch, 1987).

In some cases, perfect matching rules can be further classified as 'local' or 'nonlocal'. In this context 'local' does not refer to the way in which the rules are expressed: all matching rules are local in that sense. The distinction we wish to make refers instead to whether or not the tilings admitted by P are completely characterized by a finite atlas of unmarked patches of some finite radius. This distinction is useful in cases where it is natural to distinguish between the shapes of the prototiles and the aperiodic protoset we obtain by marking them. For example, the prototiles of tilings obtained by projection or substitution usually have very simple shapes; one wants to know which sets of prototiles can be assigned matching rules, and whether the assignment can be made by inspecting an atlas of some finite radius. More precisely, assume that we are given a set of tiles that becomes an aperiodic protoset when suitably decorated, and let T be an undecorated tiling admitted by the decorated protoset. Is there a positive integer k such that the decorations for all the tiles are unambiguously determined by their kth coronas?

Definition 7.2 Let T be a tiling whose protoset obeys a set of matching rules L. Assume that the tiles are 'unmarked'. If there is a positive integer k such that the decoration of every tile is uniquely determined by the atlas of kth coronas of T, then L is said to be local.

It follows that

Proposition 7.2 L is nonlocal if and only if the protoset admits a tiling T whose tiles can be (correctly) decorated in at least two different ways.

We have already seen that the Penrose rules are local. There are also important examples of perfect matching rules that are nonlocal. One of them is the famous Ammann octagonal tiling of the plane (Section 7.3).

Matching rules do not exist for tilings of the line: aperiodicity is possible (for repetitive tilings) only when $n > 1$. To prove this we will show that, given any repetitive nonperiodic tiling of the line with a finite protoset, it is possible to construct a periodic tiling with the same adjacencies. This means that there can be no adjacency conditions that force the tiling to be nonperiodic.

Proposition 7.3 There can be no matching rules that force nonperiodicity for repetitive tilings of the line.

Proof First we show that a vertex atlas cannot characterize a nonperiodic tiling of the line. Let the prototiles be k line segments $\alpha_1, \ldots, \alpha_k$ together with a set of rules for each of them specifying which of the k segments can be juxtaposed to it on the left, and which on the right. Consider any finite patch $\mathcal{P}$ of a tiling of the line that begins and ends with copies of the same prototile, say α_1. (Such a patch exists since the tiling is repetitive.) Repeating $\mathcal{P}$ by translation, we construct a periodic tiling that obeys the same juxtaposition rules as the nonperiodic tiling. Thus the rules do not force nonperiodicity. By a similar argument we can show that nonperiodicity cannot be forced, for repetitive tilings of the line, by specifying any finite atlas. ❑

7.2 The SCD (Schmitt–Conway–Danzer) tile

In 1988 Schmitt discovered a single tile that tiles E^3 *only* nonperiodically (Schmitt, 1988). This tile was the answer to an important question raised by Hilbert's 18th problem (Chapter 1): does there exist an aperiodic protoset with only one element? Recall that the first aperiodic protoset contained over 20 000 prototiles! But smaller examples were quickly found and by the early 1970s Penrose and Ammann had each found several aperiodic protosets of size 2. From then until Schmitt's discovery, proving the existence or nonexistence of a single aperiodic prototile was arguably one of the most important questions in tiling theory. Schmitt's construction is simple. Two parallel planes in E^3 define a slab. Mark one hyperplane with the lines of a grid, and the other with the lines of a congruent grid rotated at an angle α from the first; we assume that α is an irrational multiple of π. If superimposed, the two grids would determine a point lattice.

This point lattice is used to construct both the tile and its matching rules, in the following way. Choose a sublattice to partition the point lattice – and the slab – into congruent parallelepipedal tiles. Both the 'upper' and 'lower' faces of these tiles are automatically 'marked' with the lines of a grid. We can think of the planar strips between the lines as faces (that happen to be coplanar) or we can deform them into 'valleys' and 'mountains' (if we do this, the tiles are no longer convex). Schmitt proved that if α is suitably chosen then every tiling by direct copies of this tile (i.e. no mirror images allowed) is nonperiodic. The aperiodicity is due to the angle α, which forces the tiling to have screw rotational symmetry through an irrational angle in one direction and at the same time prevents translational symmetry in the plane orthogonal to the screw axis (because the two sets of lines, on the upper and lower faces of the tiles, are never mapped to themselves by the same translations). Since the screw rotation is an isometry of infinite order, copies of this tile appear in infinitely many orientations.

At the end of his note, Schmitt remarked that 'it may be possible to find a convex realization for the prototile, or some prototile which has the same properties and is derived from it'. Exactly such a tile was found later by Conway. Models of the tiles and two stages of building a tiling are shown in Figure 7.1; note that the tiling is not facet to facet. A net for the tile is shown in Figure 7.2. Again, no mirror images are allowed: if you join a biprism and its mirror image along a triangular facet you will obtain a single tile that admits periodic tilings of E^3.

Danzer has shown that the angles and edge lengths of Conway's tile can be so chosen, and the facets so marked, that the tiling is facet to facet in the same sense that Schmitt's was (Danzer, 1993b). In Danzer's version, the rhombic cross-section of the biprism has edge length $2n$, where n is a positive integer. Danzer proved that all of the tilings admitted by the biprism are repetitive if and only if n is odd; if n is even then it admits both

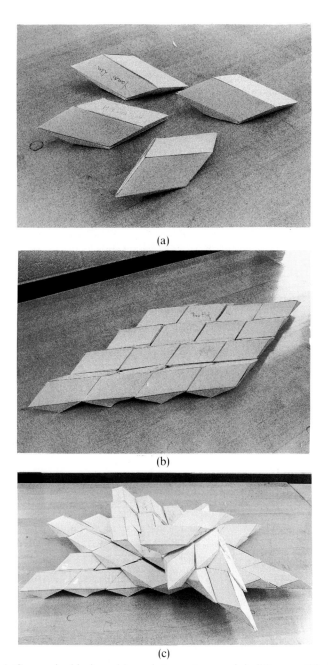

(a)

(b)

(c)

Fig. 7.1 Conway's biprism (a) and two stages of building a tiling (b, c), constructed (and signed) by participants at the National Science Foundation Regional Geometry Institute held at Smith College in July 1993.

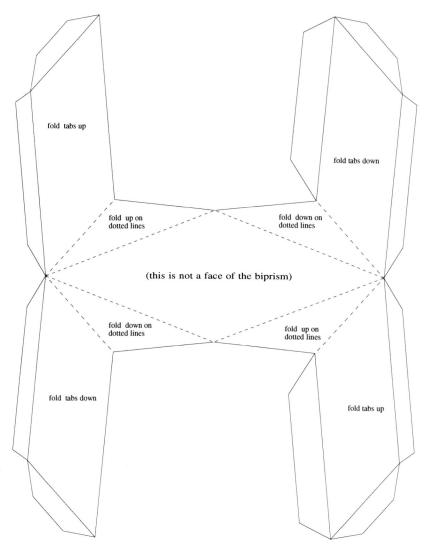

Fig. 7.2 A net for the biprism, designed by Doris Schattschneider: copy it, cut out the copy, and fold it up!

repetitive and nonrepetitive tilings. Thus when n is even the matching rule is weak.

But although the SCD tile is a monotile that seems to answer Hilbert's question, it does not seem to be 'what we were looking for', perhaps because the aperiodicity is due to a global isometry. Radin suggests that we should expand our criterion for aperiodicity to exclude any isometry – such as screw rotation – that has a translational component (Radin, 1993b).

7.3 'Octagonal' tilings

The octagonal tilings are one of a series of aperiodic tilings discovered by Ammann (all of them are illustrated in G&S and also in Ammann, Grünbaum, and Shephard, 1992). A portion of an octagonal tiling is shown in Figure 7.3. It has two prototiles: a decorated rhombus (with

Fig. 7.3 A portion of an octagonal tiling together with Ammann's prototiles.

small angle 45°) and a decorated square; the decorated square is its own mirror image, and mirror images of the rhombus are also allowed in the tiling. The name 'octagonal' reflects the fact that eight rhombs fit together at a vertex (and this configuration is globally legal).

Like the Penrose tilings, octagonal tilings can be generated both by substitution and by projection. Let us briefly look at substitution first, which as in the Penrose case, is forced by the matching rules. The mergings of the tiles of one level into the tiles of the next is shown in Figure 7.4. The associated linear map is

$$S \to 3S + 4R, \quad R \to 2S + 3R. \tag{7.1}$$

Composition is unique, so the octagonal tilings are nonperiodic by Theorem 5.2. Just as in the Penrose case, there are uncountably many of them.

The matrix

$$\begin{pmatrix} 3 & 4 \\ 2 & 3 \end{pmatrix}$$

has characteristic equation

$$\lambda^2 - 6\lambda + 1 = 0; \tag{7.2}$$

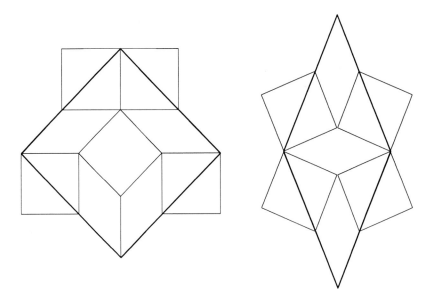

Fig. 7.4 Composition for the octagonal tilings.

its solutions are the eigenvalues $3 \pm 2\sqrt{2}$. Since

$$0 < 3 - 2\sqrt{2} < 1 < 3 + 2\sqrt{2}$$

the leading eigenvalue is a **PV** number. The relative numbers of the two kinds of tiles in the infinite tiling is the ratio of the components of the eigenvectors belonging to this eigenvalue; thus the ratio of squares to rhombs is $\sqrt{2} : 1$. This shows again that the tilings are nonperiodic.

Ammann's decorations (which express his matching rules) look much more complicated than any of those for the Penrose tiles; this is because they *are* more complicated (see Socolar, 1989). And, in fact, they are nonlocal.

There are several proofs of nonlocality (including Burkov, 1988; de Bruijn, 1989b); all are too complicated to discuss in detail here. However it is easy to see that, the Ammann tilings are not characterized by their atlas of vertex configurations (Figure 7.5), since one of these configurations tiles the plane by translation (Figure 7.6).

The nonlocality of Amman's matching rules was discovered when Beenker, a student of de Bruijn, undertook a study of 'tetragrid' tilings (a tetragrid is a k-grid, where $k = 4$; the ($\pm$) vectors of the star point to the

Fig. 7.5 The vertex atlas for the octagonal tilings.

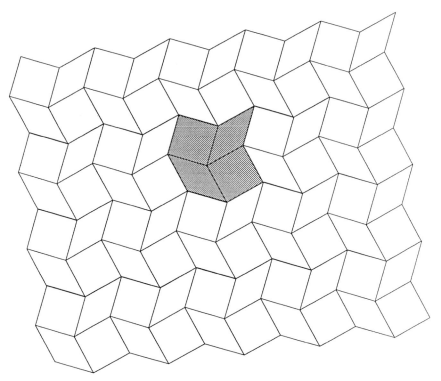

Fig. 7.6 One of the configurations of Figure 7.5 tiles the plane by translation.

vertices of a regular octagon). A portion of a tetragrid and the corresponding portion of the dual octagonal tiling are shown in Figure 7.7. As usual, the grids have been shifted so that the tetragrid is regular, that is, so that no more than two lines intersect in any point. The prototiles of the dual tiling are the unmarked rhomb and square.

A tetragrid is characterized by four shifts, which can be represented by a vector $\vec{\gamma} = (\gamma_1, \ldots, \gamma_4)$; unlike the Penrose case, there is no sum condition, but of course we require that the shift be regular. Beenker showed that any tiling dual to a regular tetragrid is a substitution tiling and, conversely, every octagonal tiling with infinitely many predecessors is the dual of a tetragrid. The Ammann octagonal tilings are substitution tilings with the same composition rule (and the same composition geometry); thus they are tetragrid duals (but Beenker was unaware of Ammann's work).

The octagonal tilings can also be obtained by canonical projection; the application of the method is straightforward. The grid star suggests that

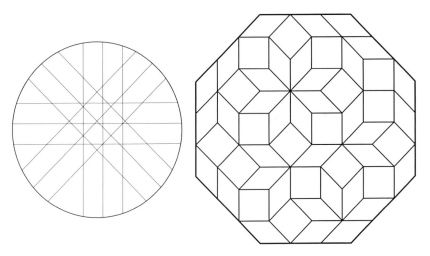

Fig. 7.7 A portion of a regular tetragrid and the corresponding portion of the dual.

the natural higher-dimensional space to work in is E^4. As we noted in Chapter 2, the symmetry group of the four-dimensional hypercube Q_4 includes a rotation of order 8. With respect to an orthonormal basis for $\mathcal{I}_4$, the matrix for this rotation can be written in the form

$$\mathcal{A} = \begin{pmatrix} 0 & 0 & 0 & -1 \\ 1 & 0 & 0 & 0 \\ 0 & 1 & 0 & 0 \\ 0 & 0 & 1 & 0 \end{pmatrix}. \tag{7.3}$$

Note that $\mathcal{A}^4 = -I$ and $\mathcal{A}^8 = I$.

The eigenvalues of $\mathcal{A}$ are the fourth roots of -1,

$$\alpha, \ -\overline{\alpha}, \ -\alpha, \ \overline{\alpha}, \qquad \alpha = \exp(\pi i/4) = \frac{\sqrt{2}}{2}(1 + i).$$

The corresponding eigenvectors are complex, but – as in the Penrose case – they span real, orthogonal planes, both of which are irrational with respect to $\mathcal{I}_4$. The isometry $\mathcal{A}$ rotates one of these planes through $\pi/4$ and the other through $3\pi/4$. We choose the first of these for the tiling plane $\mathcal{E}$. The window for the projection is an octagon.

Beenker studied the question of matching rules for tetragrid tilings by attempting to mimic de Bruijn's reconstruction of Penrose's matching rules. He showed that the edges of the squares and rhombs in the tiling could indeed be assigned arrows without contradiction, but he also found that the class B of tilings admitted by these marked prototiles is much larger than the class of tetragrid duals! Later Burkov and, independently, de Bruijn proved that *no* atlas of finite (unmarked) configurations of *any* radius completely characterizes B. In fact, Burkov has shown that for every $r > 0$, B admits periodic tilings with the same r-atlas as tetragrid tilings.

What then are we to make of Ammann's matching rules, which are perfect (in the sense of Definition 7.1)? At the time, this question seemed quite mysterious, but some of the mystery disappears when we distinguish between local and nonlocal matching rules. Ammann's rule, which restricts the tilings admitted by the prototiles to a proper subset of B, is nonlocal.

7.4 Aperiodic triangles

We have seen, in Chapter 6, that when the Penrose rhombs are bisected and properly arrowed, we obtain an aperiodic protoset consisting of four triangles. Can the tiles of every aperiodic protoset be decomposed into triangles and decorated to form a new aperiodic set? Whether or not this is the case, there are sets of aperiodic triangles other than those obtained by bisecting rhombs.

7.4.1 Danzer's triangular prototiles: seven-fold symmetry

For several years, it appeared that the symmetries possible for aperiodic tilings (in the sense of Definition 1.2) were extremely limited. In addition to the decagonal symmetries of the Penrose tilings and the icosahedral symmetry of their three-dimensional analogues, only tilings with eight-fold and twelve-fold symmetry had been found. Oddly, these were also the only rotational symmetries that had been found in real quasicrystals! It was suspected that there might be some deep connection between these phenomena, but it turned out that there were aperiodic tilings with other symmetries waiting to be discovered. Indeed, Socolar showed that weak (in the sense of Section 7.7) matching rules exist for any tiling of the plane projected from E^n when n is not a multiple of four. In the SCD tilings and the pinwheel tilings (Section 7.4.2) the tiles appear in infinitely many orientations!

An explicit example of a set of aperiodic triangle that admit tilings with seven-fold symmetry was recently found by Danzer. This work is not yet

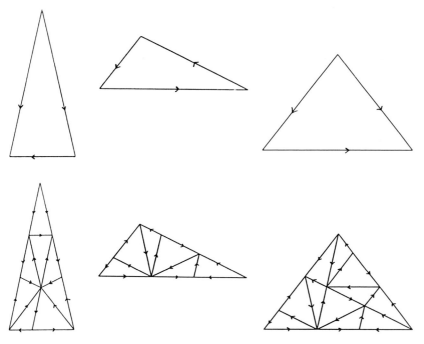

Fig. 7.8 The three prototiles of Danzer's seven-fold tiling and their substitution atlas.

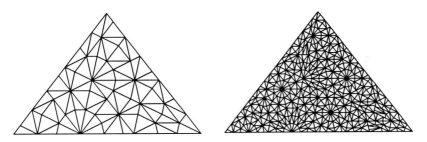

Fig. 7.9 Second and third iterations of the decomposition of the large triangle in Figure 7.8.

available in preprint form, but we report one of their results here, with the author's kind permission. The decorated prototiles are shown in Figure 7.8 together with their substitution atlas; the matching rule, which is more complicated than the atlas alone, is perfect. Second and third iterations of the decomposition of the large triangle are shown in Figure 7.9.

7.4.2 The pinwheel tilings: rotations of infinite order!

Several years ago Conway observed that a right triangle with edge lengths proportional to 1, 2, and $\sqrt{5}$ can be decomposed into five congruent copies of itself. Iterating this decomposition and rescaling, we obtain the 'pinwheel tiling' (Figure 7.10). Notice that this tiling is not edge-to-edge.

In the pinwheel tiling, triangles appear in new orientations in each generation. This is in marked contrast to the substitution tilings we have studied up till now, where the number of orientations of the tiles (in the infinite tiling) was not only finite but quite small. In the pinwheel case, the number of orientations is infinite because the smallest angle of the triangle, $\theta = \arctan(1/2)$, is irrational with respect to π.

Conway did not publish this discovery but news of it quickly spread by word of mouth. Radin has found a (very complicated) set of matching rules for these triangles that forces this composition and thus nonperiodicity (Radin, 1993a; Radin, 1994). Evidently, his matching rules are nonlocal.

7.5 Icosahedral tilings

At least three families of 'icosahedral' tilings with three-dimensional prototiles are known; we describe them briefly. We begin with a straightforward generalization of the Penrose tilings.

7.5.1 The three-dimensional Penrose tilings

Three-dimensional analogues of the Penrose tilings by decorated rhombs seem to have been introduced independently by several people; Levine and Steinhardt (1986) first drew attention to their possible role as structure models for icosahedral quasicrystals. The prototile shapes are two rhombohedra, one thick and one thin; all of their facets are congruent. Weak matching rules for these prototiles were studied by Socolar and Steinhardt (1986), based on those proposed by Ammann (unpublished). Later, Katz announced a set of perfect matching rules for them (Katz, 1988); when decorated, the number of prototiles increases from two to 22. Nets for four of these tiles are shown in Figure 7.11; the others are obtained from these by reflection and inversion. To construct the tiling, black dots must be matched to white. It is evidently not known whether these rules are local or nonlocal, or whether the tiling has the substitution property.

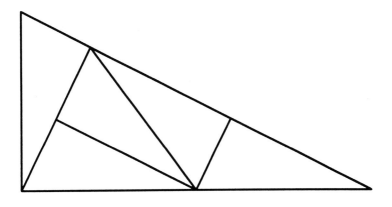

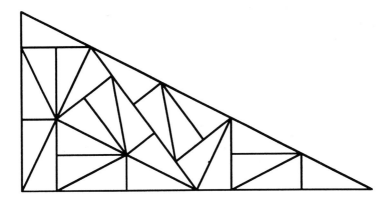

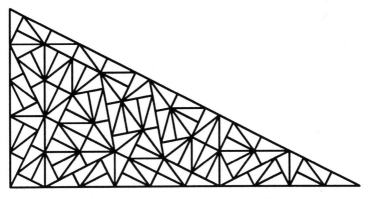

Fig. 7.10 Constructing the pinwheel tiling.

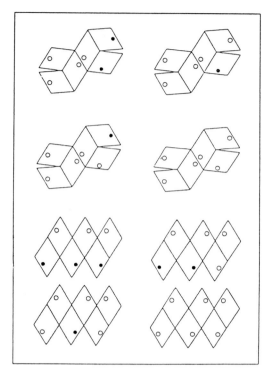

Fig. 7.11 Katz's matching rules for the 'Penrose' rhombohedra.

These three-dimensional tilings can be constructed by the standard grid method or, equivalently, by canonical projection from E^6 to E^3; it is an elaboration of the $\mathcal{I}_6$ quasicrystal discussed in Chapter 2. The frequencies of local configurations can be computed in the usual way, by studying the decomposition of the window (a triacontahedron) into subcells by its overlapping projected facets (rhombic icosahedra). See (Katz, 1989) for further details.

7.5.2 *The Socolar–Steinhardt tilings and Danzer's aperiodic tetrahedra*

Socolar and Steinhardt (1986) discovered a set of four aperiodic zonohedra: an acute rhombohedron (similar to that of Figure 7.11), a rhombic dodecahedron, a rhombic icosahedron, and a rhombic triacontahedron (notice that the second, third, and fourth of these shapes are orthogonal projections into E^3 of four-, five-, and six-dimensional hypercubes respectively). All facets are congruent and are similar to a thick Penrose rhomb; the matching rules are expressed by tile decorations. These tilings can be obtained by projection from E^6 to E^3, but the lattice used in this construction is D_6 instead of $\mathcal{I}_6$ (see Chapter 2). They can be generated by substitution as well.

Later, Danzer found a set of four aperiodic tetrahedra that build self-similar tilings of E^3 (Danzer, 1989). These tilings can be constructed by projection, but *three* windows are required!

The facets of the tetrahedra are parallel to mirror planes of a regular icosahedron I; their edges are parallel to the two-fold, three-fold and five-fold rotation axes of I and we color them red, green, and white accordingly. The dihedral angles at the red edges are $\pi/2$; at the green edges the dihedral angles are $\pi/3$ or $2\pi/3$ while the angles at the white edges are multiples of $\pi/5$.

The matching rules are:

(i) tetrahedra must share whole facets, and
(ii) facets with at least one red edge must be shared by mirror-image tetrahedra.

Composition is unique and thus the tilings are nonperiodic.
The composition matrix is

$$\mathcal{M} = \begin{pmatrix} 0 & 3 & 2 & 6 \\ 0 & 2 & 1 & 4 \\ 1 & 0 & 2 & 2 \\ 0 & 1 & 0 & 1 \end{pmatrix};$$

its eigenvalues are $\tau^3, \tau, -\tau^{-1}$ and $-\tau^{-3}$ – approximately 4.236, 1.618, −0.618, and −0.236. However τ^3 satisfies the equation $x^2 - 4x - 1 = 0$ and hence is a PV number.

It has recently been shown that the Socolar–Steinhardt and the Danzer tilings are equivalent in the sense of being 'mutually locally derivable' (Danzer, Papadopolos, and Talis, 1993); local derivability is briefly discussed in Section 7.7.

Tilings projected from E^{12} to E^3 were described by Kramer and Neri (1984).

7.6 Toward a theory of matching rules

No general theory of matching rules exists yet, but considerable progress has been made on several related subproblems. First we describe a very general method for constructing matching rules for (canonically) projected tilings. This method can be used to construct strong matching rules for a countable family of generalized Penrose tilings (the rules are perfect for a countable subset of this family). Then we will prove that every aperiodic protoset, not only those obtained by projection, admits an *uncountable* infinity of tilings. Finally we describe some recent work that shows that all matching rules (for aperiodic protosets) can lead to untileable regions.

7.6.1 Matching rules for projected tilings

All of the examples of aperiodic protosets that we have studied in this chapter (except Katz's markings for the three-dimensional rhombohedra)were found by a fortuitous combination of ingenuity, serendipity, and trial and error. But from the work of Katz, Socolar, Le, and others a theory of perfect matching rules for canonically projected is now emerging. This section describes some of the work of Le Tu Quoc Thang and his colleagues (T.Q.T. Le, S. Piunikhin, and V. Sadov, 1992a; T.Q.T. Le, S. Piunikhin, V. Sadov, 1992b; T.Q.T. Le, 1993).

The basic idea is fairly simple although its application to particular cases can be technically quite complex. Let T be a tiling obtained by canonical projection from $\mathcal{I}_n^p$ into a totally irrational d-dimensional subspace $\mathcal{E}$. The tiling T is nonperiodic but its tiles, which are unmarked d-dimensional parallelotopes, can tile $\mathcal{E}$ periodically or nonperiodically. We want to know whether there is a way to mark the prototiles so that any tiling we construct with the marked tiles will necessarily be locally isomorphic to T. Let $\mathcal{F}$ be the atlas of facet pairs of T, and let T' be any other tiling with the same atlas $\mathcal{F}$. Lift T' to a surface of d-facets in E^n. If the projections of these d-faces into $\mathcal{E}^\perp$ lie entirely in a translate of $\Pi^\perp(V(0))$, then T' must also be a canonical projection, and in fact the cylinder from which it is projected must be a translate of the cylinder for T. Thus $\mathcal{F}$ characterizes a local isomorphism class of tilings if it forces the lift to have this property. Once this is determined, one tries to find markings for the tiles that completely specify the allowed facet pairs of $\mathcal{F}$.

This problem has been studied by the 'cut' method, the alternative projection construction referred to in Chapter 5. The cut method has been developed by Kramer, Oguey, Katz, Duneau, and others. The idea is to replace the Voronoï cell of $\mathcal{I}_n^p$ by a polytope that is a union of prisms whose facets are parallel to $\mathcal{E}$ and $\mathcal{E}^\perp$. Translating this polytope by the vectors of $\mathcal{I}_n$, we obtain a periodic tiling of E^n, known as the oblique tiling. The nonperiodic tilings of $\mathcal{E}$ that we are looking for are the intersections of those translates of $\mathcal{E}$ with the oblique tilings that do not intersect the 'parallel boundary' (this term is defined below). A translate with this property is said to be regular; the singular cases can also be treated as limits of regular ones.

To show how the oblique construction works, let $n = 2$ and $d = 1$. We start with the unit square, spanned by the orthonormal basis $\vec{e}_1$ and $\vec{e}_2$ of E^2. We choose a line with irrational slope for $\mathcal{E}$; as usual, Π will be the orthogonal projector onto $\mathcal{E}$ and $\Pi^\perp$ the orthogonal projector onto $\mathcal{E}^\perp$. We define

$$\vec{\epsilon}_i = \Pi\vec{e}_i \text{ and } \vec{\epsilon}_i^\perp = -\Pi^\perp \vec{e}_i, \quad i = 1, 2.$$

Each of the pairs $\{\vec{\epsilon}_1, \vec{\epsilon}_2^\perp\}$ and $\{\vec{\epsilon}_1^\perp, \vec{\epsilon}_2\}$ defines a prism in E^2; the minus sign guarantees that the two prisms do not overlap. The union of the two prisms is again a unit cell for $\mathcal{I}_2$ (Figure 7.12); translating it by $\mathcal{I}_2$, we obtain the oblique periodic tiling $\mathcal{O}$ of E^2. (The construction can be extended to any number of dimensions.) Let T be a tile of the oblique tiling; we write

$$\partial T = T_\mathcal{E} + T_{\mathcal{E}^\perp}$$

where $T_\mathcal{E} \subset \mathcal{E}$ and $T_{\mathcal{E}^\perp} \in \mathcal{E}^\perp$.

Definition 7.3 The parallel boundary of T is the set $T_\mathcal{E}$. The parallel boundary of the oblique tiling is the union of the parallel boundaries of its tiles.

The intersection of $\mathcal{O}$ with a line parallel to $\mathcal{E}$, say $\mathcal{E} + \gamma$, where $\gamma \in E^2$ is regular, is a nonperiodic tiling of $\mathcal{E} + \gamma$ whose tiles are the intersections of $\mathcal{E} + \gamma$ with the tiles of $\mathcal{O}$ (Figure 7.13).

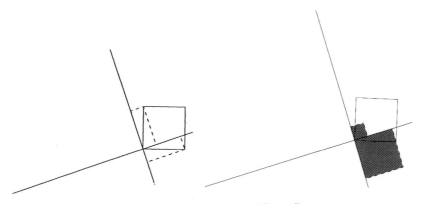

Fig. 7.12 Constructing the oblique tile.

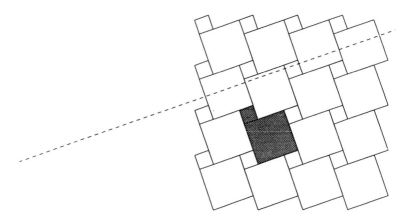

Fig. 7.13 The oblique tiling. A nonperiodic tiling of $\mathcal{E}$ is obtained by cutting the oblique tiling by a regular translate of $\mathcal{E}$ or any surface equivalent to a regular translate.

By orthogonal projection, any tiling of $\mathcal{E} + \gamma$ becomes a tiling $\mathcal{T}_\gamma$ of $\mathcal{E}$; We can obtain a tiling with the same local neighborhoods as $\mathcal{T}_\gamma$ by intersecting $\mathcal{O}$ with *any* continuous surface C (for which the restriction of Π to C is one-to-one), as long as C does not meet the parallel boundary of $\mathcal{O}$. (Crossing the parallel boundary introduces new local neighborhoods.) Such a surface is called a *section* of $\mathcal{O}$. Two sections that can be continuously deformed into one another (along $\mathcal{E}^\perp$) without meeting the parallel boundary are said to be equivalent.

To determine whether a given projected tiling $\mathcal{T}$ has matching rules, we study the lift of the tiling to the tiles of $\mathcal{O}$.

Definition 7.4 A lift of a tiling of $\mathcal{E}$ into an oblique tiling $\mathcal{O}$ is a map ϕ from the tiles of $\mathcal{E}$ to the prisms of $\mathcal{O}$ such that if T is a tile of $\mathcal{E}$, then $\Pi(\phi(T)) = T$. A lift ϕ is said to be connected if whenever T_1 and T_2 are adjacent (i.e. whenever they share a $(d-1)$-dimensional facet), then $\phi(T_1)$ and $\phi(T_2)$ are adjacent. The lift ϕ is strongly connected if $\Pi^\perp(\phi(T_1)) \cap \Pi^\perp(\phi(T_2))$ has nonempty interior.

The test for the existence of matching rules follows three steps:

  (i) ascertain whether every tiling with the same edge atlas as $\mathcal{T}$ can be lifted to a unique section of $\mathcal{O}$;
  (ii) if so, ascertain whether the lift is 'strongly connected' (i.e. whether

the projections of adjacent tiles of the lift into $\mathcal{E}^{\perp}$ have a common interior point);

(iii) determine whether the projections, into $\mathcal{E}^{\perp}$, of all finite sets of tiles of the lift have a common interior point (in which case, by Helly's theorem, the projections of *all* of the tiles do). If all three criteria are met, then every tiling with the same facet atlas as $\mathcal{T}$ belongs to the closure of the local isomorphism class of $\mathcal{T}$.

For example, we can use the oblique tiling to give another proof of Proposition 7.3 (for projected tilings). Indeed, one-dimensional tilings fail the first step, because, as Figure 7.14 shows, we can always construct a section of $\mathcal{O}$ that gives a periodic tiling. No matter how far we extend the line segments belonging to the parallel boundary (i.e. no matter how large a radius we choose for the atlas), we can still construct a periodic tiling with that same atlas. Thus one-dimensional projected tilings do not even have weak matching rules.

In cases where matching rules do exist, the facet-atlas may have to include congruent but (eventually) differently marked copies of some of its pair-configurations. One systematic way of investigating the possibilities is

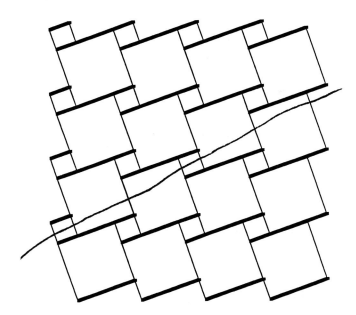

Fig. 7.14 This section defines a periodic tiling with the same vertex atlas as the tiling determined by $\mathcal{E}$.

to subdivide the prisms of $\mathcal{O}$ into smaller prisms whose boundaries are again parallel to $\mathcal{E}$ and $\mathcal{E}^{\perp}$. For bookkeeping purposes, these smaller prisms can be assigned colors; this induces a colored tiling of $\mathcal{E}$. The key theorem can be paraphrased as follows:

Theorem 7.1 Let $\mathcal{T}_c$ be a colored tiling of $\mathcal{E}$. This tiling belongs to a family obeying matching rules if it has a strongly connected lift into a colored oblique tiling that satisfies certain technical conditions.

When the conditions are satisfied, the matching rules can be obtained explicity from the atlas of colored tilings; in particular, the Penrose and Ammann matching rules are equivalent to such atlases of colored configurations. Notice that the subdivision of the prisms enlarges the parallel boundary and reduces the class of regular γs; this is how the Ammann tilings can be selected from the set of all those produced by tetragrids. In special cases (when the subdivision of the prisms of $\mathcal{O}$ does not in fact enlarge the parallel boundary) the matching rule is local; the Penrose case is one of these.

Using these methods, Le has proved:

Theorem 7.2 A generalized Penrose tiling admits perfect matching rules if and only if its parameter $\varpi = \vec{\gamma} \cdot \vec{w}$ is of the form $\varpi = (m + n\tau)/p + 1/2$, where $m, n, p \in Z$. When $p = 1$ the matching rules are local.

The (nonlocal) matching rules for the case $\varpi = 0$ (Figure 6.20b) are particularly simple and elegant (Figure 7.15); we are grateful to Le for permission to reproduce them here before publication elsewhere. For details of this construction, see (Le, 1993).

Le and Piunikhin (1993) have also proved a more general result:

Theorem 7.3 All tilings obtained by canonical projection from E^{2n} to E^n, $n \geq 2$, where E^n is a totally irrational subspace with 'quadratic slope' (that is, the basis vectors can be expressed in quadratic coordinates) admit perfect matching rules.

The rules may be local or nonlocal. This result includes Katz's rules for the three-dimensional icosahedral tilings (Section 7.5.1) and Ammann's rules for octagonal tilings of the plane (Section 7.3) as special cases.

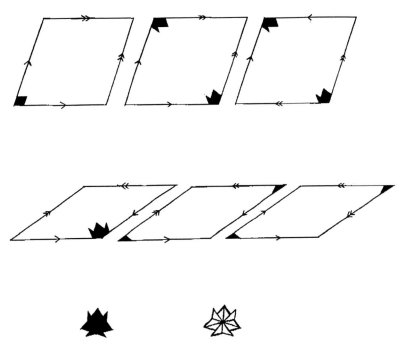

Fig. 7.15 Nonlocal matching rules for the generalized Penrose tilings of type $\varpi = 0$ (see Figure 6.20 (bottom)).

7.6.2 Uncountability is intrinsic

The aperiodic protosets that we have studied admit uncountably many tilings modulo congruence. The question naturally arises whether this is a property of *all* aperiodic protosets – except perhaps those with certain exceptional properties. The answer is yes; at least two proofs of this fact are known (Radin, private communication; Dolbilin, 1993). We give Radin's argument here. It involves some mathematical tools not called upon elsewhere in this book, but it also demonstrates the power of these tools.

Let $\mathcal{X}$ be the set of all tilings admitted by a given set of prototiles.If, in addition to the assumptions we made in Chapter 5, we assume that the boundary of a tile can be covered by other tiles in only a finite number of ways, then it can be shown (Radin and Wolff, 1992) that $\mathcal{X}$ is compact and metrizable. We fix an origin in E^2 for all the tilings of $\mathcal{X}$, and decompose $\mathcal{X}$ into orbits under translation.

Theorem 7.4 If the tilings of $\mathcal{X}$ are nonperiodic, then the number of translation orbits is uncountable.

Proof Our assumptions imply that there exists a measure μ on $\mathcal{X}$, with $\mu(\mathcal{X}) = 1$, that is both countably additive (if $X_1, X_2, \ldots$ are disjoint subsets of $\mathcal{X}$, then $\mu(\bigcup_j X_j) = \bigcup_j \mu(X_j)$) and translation invariant (for any subset $Y \subset \mathcal{X}$ and $\vec{k} \in E^n$, $\mu(Y + \vec{k}) = \mu(Y)$). The proof consists of showing that if the number of translation orbits were finite or countably infinite, then μ could not have all these properties.

Since every tiling $x \in \mathcal{X}$ is nonperiodic, every tiling $x + \vec{k}$ in the translation orbit of x is nonperiodic. If there were only countably many translation classes, then $\mathcal{X}$ could be decomposed into a countable union of disjoint translation orbits:

$$\mathcal{X} = \bigcup_j X_j,$$

where

$$X_j = \{x_j + \vec{k} : \vec{k} \in E^n\}, \quad and \quad X_j \cap X_m = \emptyset \ if \ j \neq m. \tag{7.4}$$

Since $\mu(\mathcal{X}) = \sum \mu(X_j) = 1$, there is some X_j such that

$$0 < \mu(X_j) \leq 1. \tag{7.5}$$

Now decompose E^n into half-open unit cubes; any vector $\vec{k} = (k_1, \ldots, k_n)$ lies in just one such cube. Define

$$S_{(\ell_1, \ldots, \ell_n)} = \{x_j + (k_1, \ldots, k_n) : \ell_1 \leq k_1 < \ell_1 + 1, \ldots, \ell_n \leq k_n < \ell_n + 1\}. \tag{7.6}$$

Since x_j is nonperiodic, we can decompose its translation orbit X_j into *disjoint* subsets

$$X_j = \bigcup_{(\ell_1, \ldots, \ell_n) \in Z^n} S_{(\ell_1, \ldots, \ell_n)}. \tag{7.7}$$

Since μ is translation invariant, we must have $\mu(S_{(\ell_1, \ldots, \ell_n)}) = c > 0$ for all $(\ell_1, \ldots, \ell_n) \in Z^n$. But since it is countably additive, $\mu(X_j) = \sum \mu(S_{(\ell_1, \ldots, \ell_n)})$, which contradicts $\mu(\mathcal{X}) = 1$. Thus the number of translation orbits must be uncountable.

Of course, the tilings in the different translation orbits may differ only in orientation. But the proof goes through even if we restrict $\mathcal{X}$ further in some way, for example by specifying a finite number of orientations for the prototiles. We conclude then that the prototiles must admit uncountably many *noncongruent* tilings. ❑

7.6.3 Errors are intrinsic

You have already seen, in Chapter 6, that if you try to build a Penrose tiling by juxtaposing tiles – even in accordance with the rules – you are very likely to construct an untileable region, one that cannot be corrected – within the rules – except by removing some tiles and repositioning them. To get an error-free tiling of the plane one has to iterate the substitution procedure, or construct the tilings by projection or as a pentagrid dual. These methods are global, while the matching rules act locally; this is true even when the rule is 'nonlocal' in the sense of Section 7.1. In any case, the Penrose matching rules are local in both senses of the word. It is reasonable to ask whether the 'mistake' problem can be fixed, that is, whether these rules can be strengthened to preclude the formation of untileable regions. This problem has been the subject of extensive research by physicists interested in quasicrystal growth. But they have only succeeded in constructing rules that cannot reasonably be said to be local, such as rules that require scanning arbitrarily large regions (see Lerner, 1988).

The pentagrid construction suggests that the problem may be intractable (Section 6.3), and Penrose (1989) has shown that the substitution method leads to the same conclusion. But the Penrose tilings, and these two methods, are very special cases. Is this a particular problem that arises with the Penrose tilings, or is it more general?

It turns out that the problem is very general, probably completely general. Dworkin and Shieh have recently shown that untileable regions can be constructed with *every* aperiodic protoset in the plane and (at least) a very large class of aperiodic protosets in three-dimensional space. Their proof is a generalization of the ideas used in the proof of Proposition 7.2. We will describe it for the simplest case, repetitive tilings of the plane in which copies of the prototiles appear in finitely many orientations. Choose any patch P in such a tiling. Then, because the tiling is repetitive, there exist linearly independent vectors $\vec{t}_1, \vec{t}_2$ such that $P_1 = P + \vec{t}_1$ and $P_2 = P + \vec{t}_2$ are also patches of the tiling. Let O_1 be a line segment from some point in the interior of P to some point in the interior of P_1, and O_2 a line segment from P to P_2. Let A_1 be the patch determined by P, O_1, and P_1, and A_2 the patch determined by P, O_2, and P_2. Then $B_1 \cup B_2$ is a patch, where $B_1 = P \cup A_1 \cup P_1$ and $B_2 = P \cup A_2 \cup P_2$. The patch $B_1 \cup B_2$ is shown in Figure 7.16.

Now in a new plane construct the 'parallelogram' of tiles

$$C = B_1 \cup B_2 \cup (B_1 + \vec{t}_2) \cup (B_2 + \vec{t}_1).$$

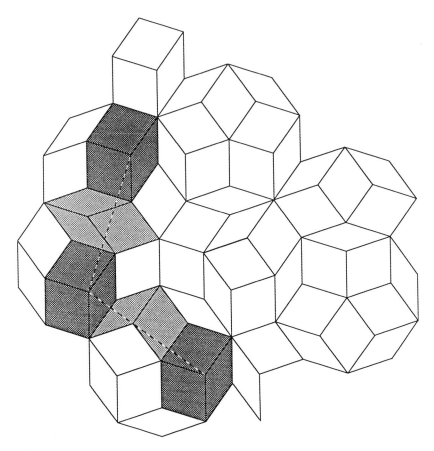

Fig. 7.16 First step in constructing an untileable region for the Penrose tiles (see text). P, P_1, P_2 are darkly shaded; A_1 and A_2 are lighter.

We can always choose $\vec{t}_1$ and $\vec{t}_2$ in such a way that this set is not itself a patch but encircles an untiled region D (Figure 7.17). Either D can be tiled (in accordance with the matching rules), or it cannot. Only the second case is possible if the protoset is aperiodic, however, because if D could be tiled, then the patch $C \cup D$ could be repeated by translation (all integral linear combinations of $\vec{t}_1$ and $\vec{t}_2$) to form a periodic tiling of the entire plane, contradicting the aperiodicity of the protoset. Thus D is untileable.

   For details of the construction in more complicated cases, for example when the tiles appear in infinitely many orientations, see Dworkin and Shieh (1993).

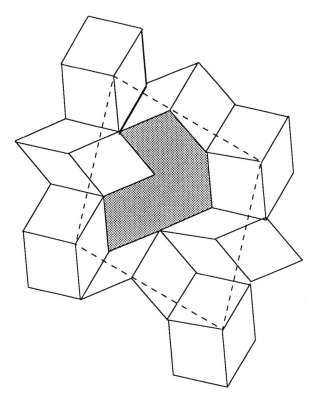

Fig. 7.17 *D* is untileable.

The near-inevitability of making 'mistakes' when building aperiodic tilings has obvious implications for quasicrystal model-building; at one time it was even thought to 'prove' that all quasicrystals must be riddled with 'defects'. The subsequent discovery of nearly defect-free quasicrystals showed, however, that in this respect at least the tiling model may be misleading. On the other hand, Socolar (1991) has shown that if we build a Penrose tiling by adding tiles, correctly, to a locally *illegal* configuration, we can ensure that further mistakes will never occur! Even periodic crystals are thought to begin growth with illegal configurations (see e.g. Senechal, 1986).

Defects in tilings and other quasicrystal models are still of interest, however, and continue to be studied (see e.g. Bohsung and Trebin, 1989). For an elegant classification of quasicrystal defects see Klèman (1992) (and also Sonneborn, 1994).

7.7 Notes

1. In the physics literature, 'strong' and 'weak' matching rules are distinguished in terms of their relation to the diffraction condition (see e.g. Ingersent, 1991; Levitov, 1988; Socolar, 1990a). For example, according to Ingersent,

Perfect matching rules are the most restrictive. Any space-filling packing constructed according to such rules necessarily belongs to a single LI class. In order for an LI class to possess perfect matching rules of range R, the R-atlas must be unique to that LI class.

Strong matching rules ensure the formation of a perfectly quasicrystalline tiling, but do not necessarily restrict the tiling to a single LI class.

Weak matching rules are the least restrictive matching rules which guarantee quasiperiodicity. They require the tiling to be at least weakly quasicrystalline.

[where]

A perfectly quasicrystalline tiling is one whose Fourier transform is composed solely of delta-function peaks arranged with a quasicrystalline – that is, crystallographically forbidden – symmetry. A weakly quasicrystalline tiling is one whose Fourier transform contains delta-function peaks arranged with quasicrystalline symmetry plus a smooth component.

In making these connections, it is evidently assumed known how these various degrees of quasicrystallinity correspond to the behavior of a lift of the tiling in an appropriate higher-dimensional space (e.g. whether its fluctuations away from an embedded copy of the tiling plane are bounded). From a mathematical standpoint, these relations remain to be proved.

2. The terminology used for matching rules in the literature can be very confusing. For example, when some authors speak of 'local rules', they mean rules that govern the way a tile is to be place in some finite patch; this is what we mean by matching rules. As we have seen, such rules may not be local in the sense of Section 7.6.1 (the distinction between local and nonlocal matching rules has only recently been clarified).

3. Planar tilings produced by 12-grids, and matching rules for them, have also received a great deal of attention; unfortunately time and space do not permit us to pay much attention to them here. For more about these tilings, see Socolar (1989), Nissen (1990), and Klitzing, Schlottmann, and Baake (1993). Socolar's aperiodic protoset, together with a portion of a (dodecagonal) tiling admitted by these prototiles, are shown in Figure 7.18.

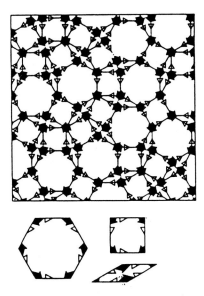

Fig. 7.18 Socolar's aperiodic protoset for a family of dodecagonal tilings, and a portion of such a tiling.

4. Evidently, many families of tilings are very closely related. For example, we saw in Chapter 6 that one can pass from a tiling by Penrose kites and darts to a tiling by thick and thin rhombs by regrouping their constituent triangles, and conversely. Other examples, such as the equivalence of the Danzer and Socolar–Steinhardt tilings, are less obvious. This suggests that tilings can (and should) be classified by 'mutual local derivability': two tilings belong to the same class if the tiles of one can be decomposed into smaller tiles which can then be regrouped to form the tiles of the other (Baake, Schottmann, and Jarvis, 1991; Schlottmann, 1993a). Mutual local derivability is clearly an equivalence relation on the set of all tilings. The existence or nonexistence of matching rules is a property of all members of a class (Klitzing, Schlottmann, and Baake, 1993; Gähler, 1994).

5. Socolar has found matching rules for the chair tiling (Figure 7.19). They have not been published elsewhere; we are grateful to him for allowing us to reproduce them here. These rules, expressed here as edge deformations, force the hierarchical structure shown in Figure 5.16.

6. Since the beginning of the nineteenth century, almost everyone has agreed that crystals grow by the accretion of building units under the influence of strictly local forces. We visualize the process as the addition of

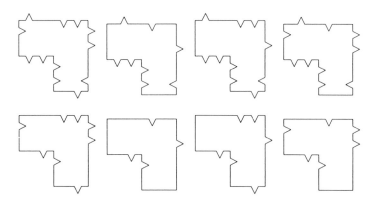

Fig. 7.19 Socolar's matching rules for the chair prototiles, expressed as edge deformations.

congruent units to a tiling of three dimensional space by bricks very much like those drawn by Haüy so long ago. One of the deepest attractions of the Penrose tiling model for quasicrystalstructure is that it permits us to retain this picture with only small modifications. But does the growth of any tiling mimic, even formally, the formation of crystalline aperiodic phases?

We should keep in mind that it is quite possible that tilings have nothing at all to teach us about quasicrystal growth. Other models of quasicrystal formation are being studied, including:

(i) Glasses. Icosahedral quasicrystals should be thought of as 'icosahedral glasses', according to Stephens (1989). An icosahedral glass is a 'system made up of clusters of atoms such that: (1) each individual cluster has icosahedral symmetry, (2) the clusters are joined so that they all have the same orientation, (3) there is some degree of randomness in the connection of the clusters, and (4) the accretion of the clusters is governed by purely local geometric rules'. Neither self-similarity nor canonical projection needs to be invoked in order to build a model satisfying these criteria. It is interesting that a structure something along these lines was initially proposed by Schectman *et al.* (1984).

(ii) Reconstructive transformations. It is possible that aperiodic crystals do not grow by accretion of either tiles or clusters: they might be modifications of periodic crystals induced by mechanical transformations (such as pressure or shear). When bonds are broken and others established in such a process, the transformation is said to be reconstructive. Dmitriev *et al.* (1990) have shown how icosahedral crystals can be constructed through a two-step process beginning with a periodic crystal with cubic atomic clusters. The clusters are first distorted to icosahedral ones, and then the structure as a whole undergoes a reconstructive transformation. While very little theoretical work on reconstructive transformations has been carried out for aperiodic crystals, there is an enormous literature about reconstructive transformations of periodic crystals. There is also a very large body of literature on 'rational approximants', the approximation of nonperiodic structures by periodic ones. Together, the transformations and approximants might lead to a powerful theory of aperiodic crystal formation, as well as some fascinating geometry.

(iii) None of the above. Penrose himself suggests that the answer to the puzzle is to be found in subtle, long-range, quantum interactions (Penrose, 1989).

8

An atlas of tiling transforms

The splitting into something discrete and something continuous
seems to me a basic issue in all morphology.

(Hermann Weyl)

To conclude this book, we return to the problem which prompted it:
trying to characterize the geometry of crystals in the broader sense of the
word. True to our promise, this problem is not solved here. We will
outline some of what is known about it today; the rest is up to you.

8.1 The atlas

We begin with an 'atlas' of tiling transforms; it is offered as a modest
supplement to *An Atlas of Optical Transforms* (Harburn, Taylor, and
Welberry, 1975), which was the inspiration for Chapter 3. Our atlas shows
small patches of nine tilings, together with computer-simulated diffraction
patterns produced by the vertices of those patches. In each plate you will
see:

(i) the patch of the tiling whose vertices constitute the (finite) set Λ of
point 'apertures' for the diffraction (there are approximately 500
vertices in each patch);

(ii) the diffraction pattern produced by these vertices (more precisely, a
gray-scale representation of the square root of the intensity
function, $\sqrt{J(\vec{s})}$);

(iii) a version of (ii) in which low-level intensities have been magnified
(the same magnification was used in all cases);

(iv) the 'negative' of (iii).

When you look at these pictures (Figures 8.1–8.9), remember that
whenever the set of apertures is finite, there will be a 'finite-size effect'.

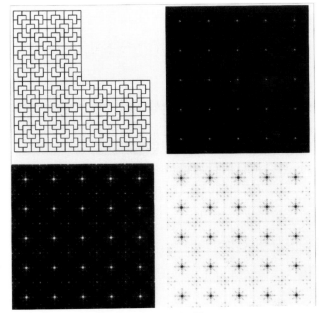

Fig. 8.1 The chair tiling.

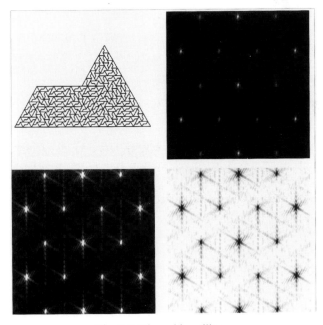

Fig. 8.2 The sphinx tiling.

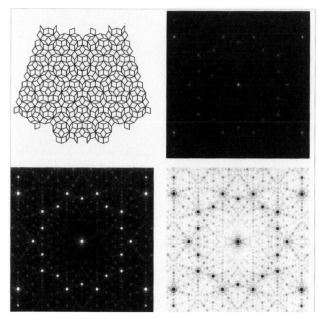

Fig. 8.3 The Penrose rhomb tiling.

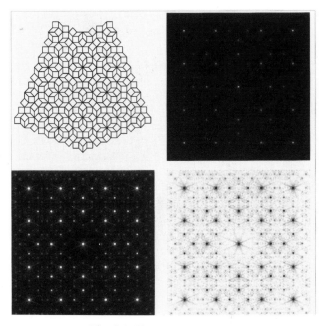

Fig. 8.4 The octagonal tiling.

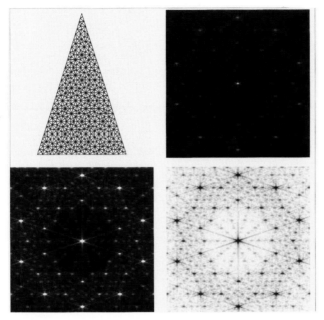

Fig. 8.5 The Penrose kite and dart tiling (with tiles bisected into triangles, as shown in Figure 6.4).

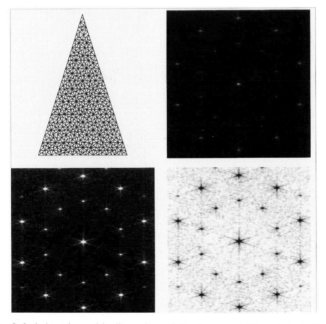

Fig. 8.6 A 'random-chiral' version of the Penrose kite and dart tiling.

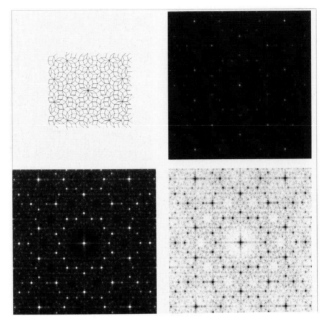

Fig. 8.7 A generalized Penrose tiling ($\varpi = 0$).

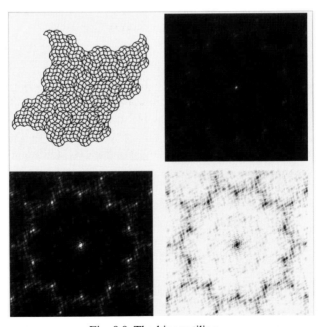

Fig. 8.8 The binary tiling.

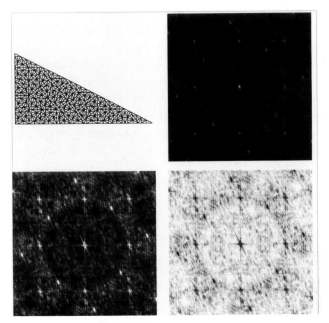

Fig. 8.9 The pinwheel tiling.

8.2 Comments

All of the tilings in our atlas, with the exception of Figure 8.7, are substitution tilings and the regions shown were generated by iterating these substitutions. Since these iterations are very greedy of computer time, we carried out only four iterations, except for the pinwheel substitution, which was iterated five times. Hopefully, this process generated sufficiently many vertices to minimize the finite-size effect and to responsibly suggest the diffraction patterns one would obtain from a tiling of the entire plane.

Let us briefly discuss the relation between the geometry of these tilings and their diffraction spectra.

8.2.1 Lattice-vacancy tilings

Although the chair and the sphinx tilings are nonperiodic, their vertices occupy sites in a lattice: the vertices of the chair tiling are a subset of the points of the square lattice, and those of the sphinx tiling are a subset of the points of a hexagonal lattice. Thus these vertex sets are crystals because the discrete components of their spectra include all the points of the reciprocal lattice (see Chapter 3). But we know that in fact the discrete component is dense in the diffraction plane, since only (full) lattices have spectra that are discrete in both senses of the word. Although at first sight these diffraction patterns appear to be periodic, close inspection of the enhanced patterns shows that this is not the case. Is there a continuous component in their spectra? We cannot say, with the methods at our disposal, but Godrèche, who used the substitution properties of the sphinx to compute its diffraction pattern, concluded that the spectrum is purely discrete (Godrèche, 1989).

8.2.2 *The Penrose and Ammann tilings*

The Penrose tilings were discussed in Chapter 6, and the Ammann tilings in Chapter 7.

Both are substitution tilings whose leading eigenvalues are PV numbers. This fact can be used to show that both of them are crystals, but the argument is a little more elaborate than the one given in Chapter 4.

Alternatively, both of these tilings can be obtained by canonical projection. In Chapter 4 we outlined three proofs that projected Fibonacci sequences are crystals: the first used the fact that these sequences have average lattices, the second made use of the dual lattice in the higher-dimensional space, and the third used Theorem 3.1 to deduce that the sequences are Poisson combs, that is, crystals with purely discrete spectra. Which of these arguments can be generalized to higher-dimensional point sets, such as the vertices of tilings of the plane? The first cannot be used unless the point set has an average lattice, which these two sets do not (but the octagonal tilings have an interesting relation to a union of *two* lattices: see Duneau and Oguey, 1990). The second argument is valid because Theorem 2.5, which depends upon uniform distribution, applies here as well. The third argument also works in any dimension and, in fact, *every* point set in E^d obtained by canonical projection from E^n is a Poisson comb.

Diffraction patterns for Penrose tilings were first studied by Mackay several years before the discovery of quasicrystals (Mackay, 1982).

8.2.3 *Generalized Penrose tilings*

Three of the tilings in this atlas can be considered to be 'generalized Penrose tilings'. The first, Figure 8.5, is a tiling by triangles, obtained by bisecting a kite and dart tiling; it is instructive to compare its diffraction pattern with that of the Penrose tiling by rhombs. As we saw in Chapter 6, the vertices of the rhomb tiling are a subset of the vertices of the kite and dart tiling. Alternatively, we can think of the set of vertices of kite and dart tiling as the union of the vertices of a rhomb tiling and another set V_p of points. We can recognize what these points are if we superimpose a rhomb tiling and its predecessor: they are the vertices of one of the families of vertex stars in the predecessor (Figure 8.10).

This diffraction pattern raises again some of the many interesting questions about what we might call crystal set theory. With respect to the predecessor rhomb tiling, $\Pi^\perp V_p$ is uniformly distributed in one of the shaded regions shown in Figure 6.23. Thus V_p is itself a crystal. This raises several question, such as whether the set of crystals is closed under finite unions and intersections, and what properties K must have in order for it to serve as the window for a projected crystal.

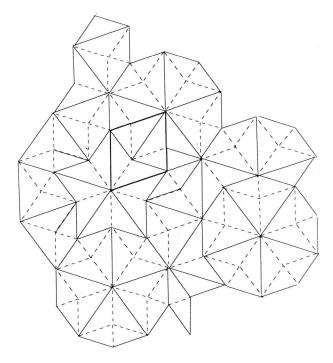

Fig. 8.10 The set of vertices of a kite and dart tiling is the union of the set V vertices of a rhomb tiling and a set V_p of points that we can recognize as the centers of a particular type of vertex star in a predecessor rhomb tiling.

The second generalized tiling, Figure 8.6, is in fact a generalization of the first. It, too, is obtained by iterating the decomposition of the kite and dart triangles, but now the correct Penrose iteration is replaced by a 'random' one, in which the decision – for each tile and at each stage – whether to decompose the triangle as if it were 'L' or 'R' is decided by a (computer) flip of the coin. Notice that the discrete component evidently persists, even though – also evidently – a continuous component now appears (see Godrèche and Luck, 1989).

The third generalized Penrose tiling is a canonical projection and thus a crystal, but this time the sum condition (Chapter 6) is disobeyed. (The fragment shown here is a small part of the region shown in Figure 6.19.) While this diffraction pattern has the same symmetry as the true Penrose rhomb tiling, it is noticeably different. The ten 'black holes' are especially intriguing.

8.2.4 A binary tiling

The binary substitution tiling can be regarded as a *very* generalized Penrose tiling. It is nonperiodic and repetitive, the shapes of its prototiles are those of the Penrose rhombs, and the relative frequency of the thick and thin rhombs is $\tau : 1$, just as in the Penrose case. But, as we noted in Chapter 6, the differences between the Penrose and binary tilings are more important than their similarities. In particular, the leading eigenvalue for the binary substitution is not a PV number. Thus although the tiling can be lifted to a connected two-dimensional surface of squares in R^5, the surface does not lie in any cylinder. This surely accounts for the hazy diffraction spectrum, but it should be emphasized that assertions in the literature that a cylinder is necessary and sufficient for 'Bragg peaks' are based on heuristic arguments, not on proved theorems. Detailed numerical analysis of the spectrum of the binary substitution tiling suggests that it is singular continuous (Godrèche, 1991; Godrèche and Lançon, 1992). Assuming that this analysis is correct, the set of vertices of this tiling is not a crystal! The tiling has been called 'quasi-quasicrystalline' by its discoverers (Lançon and Billard, 1992).

8.2.5 A pinwheel tiling

One of the most fascinating questions in quasicrystal theory is whether there exists a tiling of E^n, $n \geq 2$, with *absolutely continuous* spectrum. No example is known as of this writing, but the pinwheel tiling is a possible candidate. The symmetry of this tiling is essentially continuous, because the tiles occur in infinitely many orientations, and this will show up in the autocorrelation function of the vertex density. (The bright spots in Figure 4.9 are due to the finite size effect.) Radin has shown that there is no discrete component in its spectrum, but it is not known whether there is a singular continuous component. If it is absolutely continuous, the pinwheel tiling will be an analogue of the 'turbulent crystals' suggested by Ruelle (Ruelle, 1982; Berend and Radin, 1993).

8.3 Ordering disorder

The study of aperiodic crystals is a small part of a more general problem, that of characterizing the various degrees of what used to be called disorder (see e.g. Radin, 1991). The spectrum (of the point set or tiling) is our characterization of choice, but we still need to understand how it is related to local patterns, for example the atlases of r stars and patches.

It seems likely that, for repetitive patterns at least, the diffraction condition will turn out to be equivalent to a strong form of repetitiveness. For example, canonical projection creates tilings in which patches appear with (approximately) their global frequencies in all sufficiently large regions of the tiling, and this is just another way of saying that the projected points are uniformly distributed in the orthogonal space.

But the relations between the local and global properties of a point set or tiling are still not well understood. So instead of discussing this question inconclusively, we now describe some computer programs with which you can explore it for yourself.

8.4 Exploring with the computer

In this section we describe the computer code that we have used to draw tilings and their diffractions patterns; the code can also be obtained by anonymous ftp from minkowski.smith.edu.

8.4.1 The tilings

The substitution tilings in this atlas were drawn with the software program Mathematica, which was also used to generated the coordinates of the vertices for the c-program 'fourier' that computes the intensity function $\sqrt{J(\vec{s})}$. Our 'generic' tiling code is a modification of Wagon's code for the Penrose tilings (Wagon, 1991). It has four parts:

(i) We list the coordinates of the prototiles. From this list one can select a tile, or build a set of tiles, to be the starting configuration.

(ii) We define one or more functions called 'dissect'; they decompose the prototiles, or a list of prototiles, into copies of themselves.

(iii) We define a procedure that carries out the substitution and draws the tiles.

(iv) Finally, we define the procedure that produces the list of vertices of the tiles.

We will illustrate these steps with the easiest and most difficult examples. First, the easy one:

The chair tiling. The coordinates of our initial chair will be:

```
chair = {{2,2}, {2,4}, {0.,4}, {0.,0.}, {4,0.}, {4,2}}
```

Next, we define the function that dissects the chair and returns four smaller chairs. The order in which the parameters are listed (in the braces at the end of the routine) is crucial.

```
dissect [{ a_, b_, c_, d_, e_, f_}] := (
r = (a + b)/2; s = (d + c)/2;
t = (d + e)/2; u = (a + f)/2;
v = (3c + d + 2(a + b))/8 ;
w = (2a + 4d + c+ e )/8;
x = (2(a + f) + d + 3e)/8;
y = (2a + c + d)/4;
z = (2a + d + e)/4;
{{ a, r, v, w, x, u},{ v, r, b, c, s, y},
{ w, y, s, d, t, z}, { x, z, t, e, f, u}})
```

In order to be able to apply this function to a list of tiles, we also define:

```
dissect[list_] := Flatten[Map[dissect, list],1]
```

The function that draws the tiling is:

```
ChairDissection[n_] := (Block[{a, b, c, d, e, f}, dissect];
chairtiling = Nest[dissect, chair, n]
/. { a_, b_, c_, d_, e_, f_} ->
Line[{ a, b, c, d, e, f, a}];
Show[Graphics[chairtiling], AspectRatio -> Automatic,
PlotRange -> All])
```

```
Attributes[ChairDissection] = Listable
```

To obtain the list of vertices (of the tiles) we define:

```
Substitution[n_] := (v = chair; Do[v = dissect[v], {n}]; v)
Attributes[Substitution] = Listable
chairpoints[n_] := Union[Flatten[Substitution[n], 1]]//N.
```

The binary tiling is somewhat more complicated, in part because there are two prototiles instead of one. The computation is simplified if we think of this tiling as lying in the complex plane.

First we define some constants that we will need:

```
t = GoldenRatio//N
a = (Cos[Pi/10] + I Sin[Pi/10])//N
b = (Cos[2 Pi/5] + I Sin[2 Pi/5])//N
c = (Cos[ Pi/5] + I Sin[ Pi/5])//N
d = Sqrt[2 + t]
```

Next we give the coordinates of the two prototiles. The fifth coordinate, ± 1, serves to distinguish the thick and thin rhombs; it will be dropped later:

```
thick = {0., 1.5, 1.5 + 1.5 b, 1.5 b, 1}
thin = {0, 1.5, 1.5 + 1.5 c, 1.5 c, -1}
```

Now we define dissection for each of the prototiles, and for a list of prototiles:

```
dissect[{ p_, q_, r_, s_, -1}] := (
w = p + ((1/d) (q - p)) Conjugate[a];
y = p + ((1/d) (s - p)) a;
z = w + y -p;
u = z + s- y;
{{ z, y, p, w, 1},
{ w, q, u, z, -1}, { u, s, y, z,-1}})

dissect[{ p_, q_, r_, s_, 1}] := (
w = p + (1/d) (q-p) Conjugate[a];
y = p + (1/d) (s - p) Conjugate[a];
```

```
z = w + y -p;
u = z + s -y ;
v = q + z-w;
 {{ z, y, p, w, 1}, { z, u, s, y, 1},
 { z, v, r, u, 1}, { w, q, v, z, -1}})

dissect[list_] := Map[dissect, list]
```

Here is the function that draws the tiling, beginning with a thin rhomb. (To start with a thick rhomb, just change 'thin' to 'thick' in the code.)

```
Config[n_] := (
tiles = Nest[dissect, thin, n] /.
{p_, q_, r_, s_, x_} ->
Line[{{ Re[p], Im[p]},
{ Re[q], Im[q]},
{ Re[r], Im[r]},
{ Re[s], Im[s]},
{ Re[p], Im[p]}}];
Show[Graphics[tiles], AspectRatio-> Automatic,
PlotRange -> All] )
```

To obtain the list of vertices, we define a function that drops the last element of the list on which it operates:

```
dropminus1[list_] := Drop[list, -1]

Substitution[n_] := (
v = thin; Do[v = dissect[v], {n}];
v = Flatten[v, n-1];
v = Map[dropminus1, v];
v = Flatten[v, 1];
Table[{Re[v[[j]]], Im[v[[j]]]}, {j, 1, Length[v]}])

Attributes[Substitution] = Listable
binarypoints[n_] := Union[Substitution[n]]//N
```

The output of 'chairpoints' and 'binarypoints' can be piped to files for use in 'fourier' (but they must first be stripped of all brackets and commas).

8.4.2 'fourier'.c

This c-program, written by Stuart Levy, inputs a list of coordinates of points in the plane, computes the square root of the intensity function for these points, and produces portable gray-map image files.

```c
#include <stdio.h>
#include <stdlib.h>
#include <math.h>    /* Declares sin(), cos(), etc. */


/*
 * To compile:
 * cc -o fourier fourier.c -lm
 */


struct point {
double x, y;
};


int
getnumber(double *v)
{
int c;
  for(;;) {
switch(c = getchar()) {
case ' ': case '\t': case '\n': case'\r':
case ',': case '(': case ')': case '{': case '}': case '[': case ']':
continue; /* Ignore white space and Mathematica punctuation */

  case '#': /* # denotes comments: skip to end of line. */

while((c = getchar()) != '\n' && c != EOF)
;
continue;

  default:
ungetc(c, stdin);
return (scanf("%lf", v) == 1);
}
}
}
```

```
main(argc, argv)

char *argv[]; {
double srange = 10.0;
int n = 100;
double brightness = 1;
double scale;
int i, j, k;
struct point s;
struct point *source;
double re, im;
int nsource, allocated;
double dot, mag, maxmag;

if(argc <= 1 || (n = atoi(argv[1])) <= 0) {
fprintf(stderr, "\

Usage: fourier N srange [brightness] <point_position_file >
  image_file.pgm\n\
Computes the Fourier transform of a collection of points; produces
an NxN image\n\covering the range (-srange,-srange) to (srange,srange)
in Fourier space.\n\n\Reads (x,y) point positions from standard
input as a list of floating-point\n\numbers separated by blanks.
Writes a PGM-format image to standard output.\n\The optional
\"brightness\" (default 1) brightens the image by the given factor;\n\
the default range of output values is chosen to prevent saturation
in the worst case.\n");
exit(1);
}

if(n == 1) n = 2;
if(argc > 2) srange = atof(argv[2]);
if(argc > 3) brightness = atof(argv[3]);

/*
* Read light sources. Enlarge the table when necessary.
*/
allocated = 15;
source = (struct point *)malloc(allocated * sizeof(struct point));
```

```
nsource = 0;
while(getnumber(&source[nsource].x) && getnumber
  (&source[nsource].y))
{
nsource++;
if(nsource >= allocated) {
allocated *= 2;
source = (struct point *)realloc(source,
allocated * sizeof(struct point));
}
}
fprintf(stderr, "%d light sources.\n", nsource);

scale = nsource > 0 ? brightness * 256 / nsource : 1;

/*
 * Produce portable-graymap header
 */
printf("P5\n%d %d\n255\n", n, n);

maxmag = 0;
for(i = 0; i < n; i++) {
s.y = srange*((float)i/(n-1) - .5);

for(j = 0; j < n; j++) {
s.x = srange*((float)j/(n-1) - .5);

re = im = 0;
for(k = 0; k < nsource; k++) {
dot = 2*M_PI*(s.x*source[k].x + s.y*source[k].y);
re += cos(dot);
im += sin(dot);
}

mag = scale * sqrt(re*re + im*im);
putchar(mag >= 255 ? 255 : (int)mag);
if(maxmag < mag) maxmag = mag;
}
}
if(maxmag > 0)
fprintf(stderr, "Max mag %.1f; max brightness which won't saturate:
```

```
%g\n",
maxmag, brightness * 256/maxmag);
exit(0);
}


}
```

8.4.3 *Line arrangements and zonotopal tilings*

We thank Jürgen Richter-Gebert and Kajo Miyazaki for permission to reproduce here the postscript code that draws line arrangements, including portions of multigrids and the tilings dual to them, such as Figures 5.19 and 7.7 (Richter-Gebert, 1993).

```
!PS-Adobe-2.0 EPSF-2.0
/* Title: example4.eps
/* Creator: JRG
/* Pages: 0 1
BoundingBox: 0 0 600 800
EndComments
/*BeginProcSet: arrows 1.0 0

300 570 translate
0.7 0.7 scale
0 setgray
/defvar {exch def} def
/reg {/win defvar win cos win sin} def
/gon {div 180 mul reg} def
/ssc {sss mul} def
/sss 1 def

% Here start the definition of the line data

/Lines [
/* Homogous Coordinates (x,y,d),
/* (x,y) shoud be the normalVector,
/* d should be the distance from the origin:
/* The lines must be sorted by increasing slope.
```

```
/n 5 def
/* x:   y:   d:
0 1 n 1 sub {
/winkel defvar

winkel n gon 0.26 ssc
winkel n gon 0.16 ssc
winkel n gon 0.06 ssc
winkel n gon −0.04 ssc
winkel n gon −0.14 ssc
winkel n gon −0.24 ssc
/sss sss neg def
} for

] def

% Here start many useful functions and definitions

/size 70 def

/NumberOfLines Lines length 3 div .1 add cvi def

/LineWithIndex

{/lineIndex defvar
Lines lineIndex 3 mul get
Lines lineIndex 3 mul 1 add get
Lines lineIndex 3 mul 2 add get } def

/DirectionWithIndex
{LineWithIndex pop } def

/LineScale
{/ss defvar
ss mul exch ss mul exch} def

/MovePositiveInDirection
{DirectionWithIndex 1 size mul LineScale rmoveto} def

/MoveNegativeInDirection
{DirectionWithIndex -1 size mul LineScale rmoveto} def

/LinePositiveInDirection
{DirectionWithIndex 1 size mul LineScale rlineto} def

/LineNegativeInDirection
```

```
{DirectionWithIndex -1 size mul LineScale rlineto} def

/Det { /m11 defvar /m12 defvar /m13 defvar
/m21 defvar /m22 defvar /m23 defvar
/m31 defvar /m32 defvar /m33 defvar


m11 m22 m33 mul mul
m12 m23 m31 mul mul add
m13 m21 m32 mul mul add
m11 m23 m32 mul mul neg add
m12 m21 m33 mul mul neg add
m13 m22 m31 mul mul neg add
} def


/Sign {dup .00001 ge {pop 1} {-.00001 le {-1} {0} ifelse} ifelse} def


/SignVector{
[
0 1 NumberOfLines 1 sub {
LineWithIndex
index1 LineWithIndex
index2 LineWithIndex Det Sign } for]
} def

/DrawTile {
/cv defvar
0 0 moveto
0 1 NumberOfLines 1 sub {
/index defvar
/sign {cv index get} def


1 sign eq {index MovePositiveInDirection}
{index MoveNegativeInDirection} ifelse
} for


0 1 NumberOfLines 1 sub {
/index defvar
/sign {cv index get} def
0 sign eq {index LinePositiveInDirection
index LinePositiveInDirection} if
} for
```

```
0 1 NumberOfLines 1 sub {
/index defvar
/sign {cv index get} def
0 sign eq {index LineNegativeInDirection
index LineNegativeInDirection} if
} for
} def

% Here starts the drawing of the line arrangement

gsave

0 −500 translate
1.3 1.3 scale
1.3 setlinewidth
0 0 150 0 360 arc stroke
0 0 150 0 360 arc clip
0.7 setlinewidth

0 1 NumberOfLines 1 sub {
LineWithIndex
gsave
3 mul size mul /distance defvar
atan rotate
−10000 distance moveto 10000 distance lineto stroke
grestore
} for

grestore

% Here starts the drawing of the zonozopal tiling

NumberOfLines 3 div setlinewidth
6 NumberOfLines div 6 NumberOfLines div scale
1 setlinejoin
[1 0 0 -1 0 0] concat
0 1 NumberOfLines 2 sub {/index1 defvar
index1 1 add 1 NumberOfLines 1 sub {/index2 defvar
SignVector DrawTile} for } for stroke

NumberOfLines 2 div setlinewidth
[0 1 NumberOfLines 1 sub{pop 0} for] DrawTile stroke
```

```
showpage
}
```

8.4.4 QuasiTiler

The computer program 'QuasiTiler', by Eugenio Durand, runs on NeXTStep software and can be obtained by anonymous ftp from the Geometry Center (ftp geom.umn.edu). It will draw (almost) any desired projected tiling, including the Penrose tilings and the generalized Penrose tilings obtained by changing the value of the parameter ϖ (see Chapter 6). Some of the many features of this program can also be enjoyed via Mosaic and the World Wide Web.

Appendix I

A mathematical toolbag

A. Numbers

1. Notation. We will use the following symbols:

N, the set of nonnegative integers

Z, the set of integers

Q, the set of rational numbers

R, the set of real numbers

C, the set of complex numbers $a + bi$, $a, b \in R$, $i^2 = -1$

Clearly, $N \subset Z \subset Q \subset R \subset C$.

We write $2Z$ for the set of all even integers, $3Z$ for the set of all multiples of three, and so on.

Z^n will denote the set of all n-tuples of integers and also the free abelian group with n generators. R^n will denote the set of all n-tuples of real numbers and also n-dimensional affine space.

E^n is n-dimensional Euclidean space, that is, R^n together with the Euclidean metric:

$$|(x_1, \ldots, x_n)|^2 = x_1^2 + \cdots x_n^2.$$

2. Classification of real numbers. Real numbers which are not rational are *irrational*. Among the irrational numbers, we distinguish the algebraic and the transcendental.

A real number is said to be *algebraic* if it is the root of a monic polynomial equation with rational coefficients ('monic' means that the leading coefficient is equal to 1). For example, the golden number $\tau = (1 + \sqrt{5})/2$ is algebraic because it is a root of the equation

$$x^2 - x - 1 = 0.$$

A real number that is not algebraic is *transcendental*; π and e are the most famous transcendental numbers.

Every algebraic number ω satisfies infinitely many polynomial equations. For example, in addition to satisfying $x - 2 = 0$, 2 is the solution of $2x - 4 = 0$ and a solution of $x^2 + x - 6 = 0$. But among all these polynomials there is exactly one monic polynomial whose degree is minimal. This is the *minimal polynomial* of ω. The *degree* of ω is the degree of its minimal polynomial. A polynomial of degree n has n roots (real or complex, and not necessarily different from one another). The roots of the minimal polynomial of ω are its *conjugates*.

When the coefficients of its minimal polynomial are integers, the algebraic number is said to be an *algebraic integer* (thus τ is an algebraic integer). An algebraic integer is a *PV* (Pisot–Vijayvaraghavan) number if it is greater than 1 and all its conjugates are strictly less than one in absolute value. Among the algebraic integers the *PV* numbers play a crucial role in quasicrystal theory.

3. Infinities. A set whose elements can be placed in one-to-one correspondence with the positive integers is said to be *countable*; for convenience we also say that finite sets are countable. Sets which are not countable are *uncountable*. Q is countable, but R is uncountable. The algebraic numbers are countable; thus almost all real numbers are transcendental.

B. Eigenvalues, eigenvectors, and eigenspaces

Any linear transformation T of an n-dimensional vector space V can be represented by a conjugation class of matrices, each of which describes the action of T on E^n with respect to different choices of bases for V. The simplest description is one in which the basis vectors are disturbed as little as possible by T. The simplest thing that a transformation can do to a vector is of course to leave it fixed; the next simplest thing is to keep it on the line on which it lies. Thus we are looking for nonzero vectors $\vec{x}$ such that

$$T\vec{x} = \lambda\vec{x}, \tag{AI.1}$$

where λ is a scalar. λ is called an *eigenvalue* of T and $\vec{x}$ is an *eigenvector* belonging to λ. Notice that if $\vec{x}$ satisfies (AI.1), then so does $\alpha\vec{x}$ for any scalar α, since T is linear.

If $\varepsilon_1, \ldots, \varepsilon_k$ are eigenvectors of T belonging to the same eigenvalue, then the k-dimensional subspace they span is stabilized by T. Such a subspace is called an eigenspace.

It is easier to find the eigenvalues λ first and then the eigenvectors belonging to them, rather than the other way around. Equation (AI.1) can be rewritten in the form

$$\left(T - \lambda I\right)\vec{x} = 0,$$

where I is the identity matrix. Since we are looking for nonzero $\vec{x}$, we are looking for values of λ for which

$$\det(T - \lambda I) = 0. \tag{AI.2}$$

Equation (AI.2) is a polynomial equation of degree n, called the characteristic polynomial of T; its solutions are the eigenvalues of T. When T can be represented by an integral matrix, then its characteristic polynomial has integer coefficients and the leading coefficient is equal to one. Thus the eigenvalues are algebraic integers.

In quasicrystal theory, it is important to distinguish between left and right eigenvectors. Let λ be an eigenvalue of T, and let A be any matrix representation of T. If $\vec{x}$ is a row vector, then $\vec{x}^\top$ is the corresponding column vector.

Definition AI.1 A left eigenvector of T is a vector $\vec{x}$ such that $\vec{x}A = \lambda\vec{x}$. A right eigenvector is a vector $\vec{x}$ such that $A\vec{x}^\top = \lambda\vec{x}^\top$.

The matrices we deal with in the theory of substitution tilings are *primitive*: all of their entries are nonnegative integers, and for some power all entries are positive.

The crucial property of primitive matrices is based on the famous

Theorem AI.1 (Perron–Frobenius) Every matrix with positive entries has a real eigenvalue w_1 such that $|w_1| > |w_j|, j = 2, \ldots, k$, where $w_2, \ldots, w_k$ are the algebraic conjugates of w_1.

The primitive matrices 'count' the tiles in successive patches of a tiling (generated by substitution) and give the relative sizes of the tiles in a self-similar tiling family; it is here that the distinction between left and right multiplication is crucial. For example, let the substitution rule be

$$A' = \alpha A + \beta B, \quad B' = \gamma A + \delta B,$$

where the matrix

$$M = \begin{pmatrix} \alpha & \beta \\ \gamma & \delta \end{pmatrix}$$

is primitive. If we generate patches beginning, say, with one A tile, that is, an initial population vector $\vec{p}_0 = (1, 0)$, then in the first generation we will have α A tiles and β B tiles, which we obtain by multiplying M on the *left*:

$$(\alpha, \beta) = (1, 0) \begin{pmatrix} \alpha & \beta \\ \gamma & \delta \end{pmatrix}.$$

On the other hand, if we require that the relative ratios of large to small tiles is preserved under substitution, then we must have

$$\frac{\text{vol } A'}{\text{vol } B'} = \frac{\text{vol } A}{\text{vol } B}$$

or

$$\frac{\alpha \text{ vol } A + \beta \text{ vol } B}{\gamma \text{ vol } A + \delta \text{ vol } B} = \frac{\text{vol } A}{\text{vol } B}.$$

If you work this out you will find that $(\text{vol } A, \text{vol } B)$ must be a *right* eigenvector of M.

C. Group action

Let G be a transformation group acting on a 'mathematical object' M. (M may be all of E^n, or a geometrical object such as a cube, or even G itself.) Let x be a point of M. For our purposes, 'group action' can be thought of as the study of what happens to x under the transformations in G.

Definition AI.2 The orbit of x under G, is the set

$$O_G(x) = \{gx \in M | g \in G\}.$$

Definition AI.3 The stabilizer (or isotropy) group of x is the maximal subgroup of G for which x is a stable set.

The following facts can easily be proved by elementary group theory.

(i) The orbits of two points of M, say x and y, are either identical or disjoint:

$$y \in O_G(x) \iff O_G(y) = O_G(x).$$

(ii)   Points in the same orbit have conjugate stabilizers: there is a $g \in G$ such that $gy = x$ if and only if

$$S_G(y) = g^{-1}S_G(x)g.$$

Points with conjugate stabilizers need not be in the same orbit (i.e., if they are conjugate by an outer automorphism).

D. Number-theoretic functions

Only two are used in this book.

1. The greatest integer function and its relatives. Every real number x can be written as the sum of an integer and a number in the half-open interval $[0,1)$. The integer part of x is the *greatest integer less than or equal to x* and is denoted by $[x]$; the number lying in $[0,1)$ is the *fractional part of x* and is denoted by $\{x\}$. Thus

$$x = [x] + \{x\} \ .$$

The function $y = [x]$ is a step function; the function $y = \{x\}$ is periodic. Their graphs are shown in Figure AI.1.

The greatest integer function is also sometimes (but not in this book) called the *floor* function and is denoted by $\lfloor x \rfloor$. Correspondingly, there is the *roof* function $\lceil x \rceil$, the smallest integer greater than or equal to x, defined by

$$\lceil x \rceil = \begin{cases} \lfloor x \rfloor + 1, & \text{if } x \notin Z, \\ x, & \text{if } x \in Z. \end{cases}$$

We will not use floor and roof, but we will use the *nearest integer function*, $\|x\|$. If $m \in Z$ and $m \le x < m + 1$, then

$$\|x\| = \begin{cases} m & \text{if } x \in [m, m + 1/2) \\ m + 1 & \text{if } x \in [m + 1/2, m+1). \end{cases}$$

We set $\|x\| = m + 1$ if $x = m + 1/2$. Then $\|x\|$ has the simple definition

$$\|x\| = [x + \tfrac{1}{2}].$$

We can also write

$$\|x\| = x + \frac{1}{2} - \{x + \frac{1}{2}\}.$$

2. Euler's ϕ-function. This function is defined on the positive integers: $\phi(n)$ is the number of (positive) integers less than and relatively prime to n. Thus $\phi(5) = 4$, $\phi(6) = 2$, and so forth. (We set $\phi(1) = 1$.) The values of ϕ can easily be computed because it has the following two properties:

(i)   if n is a power of a single prime, that is, if $n = p^k$, where p is a prime and k a positive integer, then $\phi(p^k) = p^{k-1}(p - 1)$;
(ii)   if n_1 and n_2 are relatively prime, then $\phi(n_1 n_2) = \phi(n_1)\phi(n_2)$. For example, $\phi(100) = \phi(2^2 5^2) = \phi(2^2)\phi(5^2) = 2 \times 20 = 40$.

3. Big O and little o. Although these symbols are not really number theoretic functions, they are frequently used in number theory (as well as in other fields of mathematics). Let $f(x)$ and $g(x) > 0$ be two functions defined for all x. We say

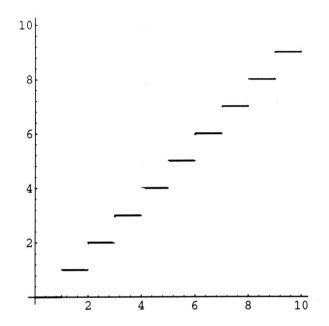

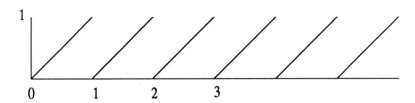

Fig. AI.1 (Top) The graph of $y = [x]$. (Bottom) The graph of $y = \{x\}$.

that

$$f(x) = O(g(x)) \quad \text{if} \quad \lim_{x \to \infty} \frac{|f(x)|}{g(x)} < \infty,$$

and

$$f(x) = o(g(x)) \quad \text{if} \quad \lim_{x \to \infty} \frac{|f(x)|}{g(x)} = 0.$$

E. Density and uniform distribution

A set A is said to be dense in a set B if, for every b in B and for all $\epsilon > 0$, every ball $B_b(\epsilon)$ contains a point of A. For example, Q is dense in R but Z is not.

A simple but important example of a set dense in the open unit interval $(0,1)$ is the sequence of fractional parts $\{n\theta\}$ where θ is irrational. Density follows immediately from Kronecker's theorem, which in one of its many forms, says:

Theorem AI.2 (Kronecker). If θ is irrational and $\alpha \in (0,1)$ then for every positive integer N and $\epsilon > 0$ there is an $n > N$ such that $|\{n\theta\} - \alpha| < \epsilon$.

There are many generalizations of this theorem to higher dimensions; see, for example, Hardy and Wright (1962).

The sequence $\{n\theta\}$ is not only dense in the unit interval, it is uniformly distributed there. Let $x = (x_n)$ be an infinite sequence of real numbers in $(0,1)$, let J be any subinterval of $(0,1)$, and let $N_J(x)$ be the number of terms of x, among the first N, whose fractional parts lie in J. The sequence x is said to be *uniformly distributed* in the interval $(0,1)$ if

$$\lim_{N \to \infty} N_J(x) = |J|.$$

For example, if N is sufficiently large then approximately half of the numbers $x_1, \ldots, x_N$ will lie in the interval $(0,1/2)$.

To appreciate the subtlety of uniform distribution we consider the following problem. Let f be a continuous function defined on the closed unit interval. The Riemann integral $\int_0^1 f(x)dx$ is the limit of the sequence of partial sums:

$$\lim_{N \to \infty} \sum_{n=1}^{N} f(x_n)\frac{1}{N} = \int_0^1 f(x)dx, \tag{AI.3}$$

where, for each N, x_n is chosen to be some point in the interval $((n-1)/N, n/N)$. Suppose now that instead of freely selecting the x_n after each choice of N, they are required to be the first N terms of a given sequence. What conditions must the sequence satisfy in order for the sequence of partial sums in (AI.3) to converge to the integral?

Theorem AI.3 (Weyl) The following statements are equivalent:

(i) The sequence (x_n) is uniformly distributed in the unit interval;

(ii)
$$\lim_{N \to \infty} \frac{1}{N} \sum_{n=1}^{N} f(x_n) = \int_0^1 f(x)dx$$

for every continuous function – possibly complex-valued – defined on $(0,1)$;

(iii)
$$\lim_{N \to \infty} \frac{1}{N} \sum_{n=1}^{N} \exp(2\pi i m x_n) = 0 \tag{AI.4}$$

for every nonzero $m \in Z$.

For a proof, see Weyl (1916) or Cassells (1965).

It follows immediately from (iii) that the sequence $\{n\theta\}$ is uniformly distributed in $(0,1)$ because the sum in (AI.4) is a convergent geometric series.

It is important to realize that uniform distribution is defined by a *process*: it depends on the ordering of the terms in a sequence. It is meaningless to ask whether a sequence was uniformly distributed after the fact. Indeed, a theorem of von Neumann states that *the terms of any sequence dense in* $(0,1)$ *can be rearranged into a uniformly distributed sequence* (see e.g. Hlawka 1979).

F. The structure of orbits of Z-modules. Theorem 2.3 is closely related to a classical theorem on the closed subgroups of R^n (see, for example, Bourbaki,

1966). This formulation of the theorem and its proof were contributed by R. Moody.

Theorem A1.4 Let $\phi : R^n \to R^m$ be a surjective linear mapping, where $m < n$. Then there are subspaces V, W of R^m such that $R^m = V \oplus W$ and

   (i) $\phi(Z^n) = \phi(Z^n) \cap W + \phi(Z^n) \cap V$,
  (ii) $\phi(Z^n) \cap W$ is a discrete lattice in W, and
 (iii) $\phi(Z^n) \cap V$ is a dense subgroup of V.

In particular $\phi(Z^n) \simeq L \times M$ (discrete $\times$ dense).

Proof Let us assume that $\phi(Z^n) \simeq Z^n$ has an accumulation point. We will show that in this case the subspace V is nontrivial.

Since $\phi(Z^n)$ is a group, 0 is an accumulation point. For each $r > 0$ and $v \in R^m$, let $U_r(v)$ be the open ball of radius r about v. Since 0 is an accumulation point, we have, for all $r > 0, \{0\} \subset U_r(0) \cap \phi(Z^n)$, the inclusion being proper. Let $V_r :=$ the R-linear span of $U_r(0) \cap \phi(Z^n)$. Then $V_r \neq 0$ and $r > s$ implies $V_s \subset V_r$. Set

$$V = \cap_{r>0} V_r.$$

For all sufficiently small r, $V_r = V$ and $U_r(0) \cap \phi(Z^n)$ spans V.

Let

$$K := \phi^{-1}(V) \cap Z^n = \{x \in Z^n | \phi(x) \in V\}.$$

K is a subgroup of Z^n and hence there is a basis $e_1, \ldots, e_n$ of Z^n and positive integers $a_1, \ldots, a_k$ for some $k \leq n$ such that $\{a_1 e_1, \ldots, a_k e_k\}$ is a basis of K. By the definition of K we see that $a_i e_i \in K$ implies $e_k \in K$, so

$$K = Ze_1 \oplus \cdots \oplus Ze_k.$$

We define

$$L := Ze_{k+1} + \cdots Ze_n$$

so $Z^n = L \oplus K$.

Let $\pi : R^m \to R^m/V$ be the natural map and let $\eta : L \to R^m/V$ be $\eta = \pi \circ (\phi|_L)$. Then

   (i) η is injective since $L \cap \phi^{-1}(V) = 0$,
  (ii) $\eta(L)$ spans R^m/V since $\phi(Z^n)$ spans R^m,
 (iii) $\eta(L)$ is discrete.

In fact, suppose that $\{\ell_i\}$ is a sequence of elements of L such that $\{\eta(\ell_i)\}$ converges in R^m/V. As usual we can assume that $\{\eta(\ell_i)\}$ converges to 0. Write $R^m = V' \oplus V$ and

$$\phi(\ell_i) = v'_i + v_i, \ \forall i, \ v'_i \in V', v_i \in V.$$

We can identify R^m/V and V'. The hypothesis is then that $\{v'_i\} \to 0$. Since $\phi(K)$ is dense in V one can choose $k_i \in K$ such that $|v_i - \phi(k_i)| < 1/(2i)$. Then

$$|\phi(\ell_i - k_i)| = |\phi(\ell_i) - \phi(k_i)| = |v'_i + v_i - \phi(k_i)|$$
$$\leq |v'_i| + |v_i - \phi(k_i)| \to 0.$$

Thus for positive i, $\phi(\ell_i - k_i) \in V$ and $\ell_i - k_i \in K$. Then $\ell_i \in K \cap L = 0$ and $\eta(\ell_i) = 0$. This proves discreteness.

From (ii) and (iii), $\eta(L)$ is a lattice of rank = $\dim R^m - \dim V$. From (i), L has the same rank. Thus $n - k = m - \dim V$. Define W to be the real span of $\phi(L)$. Since $L = Ze_{k+1} + \cdots + Ze_n$, $\dim W \le n - k = m - \dim V$ and since, clearly, $R^m = W + V$ we have $\dim W = m - \dim V$ and $R^m = W \oplus V$. Since $\eta(L)$ is discrete and π is continuous, $\phi(L)$ is discrete and also has the same rank as L. So $L \simeq \phi(L)$ and $\phi(L)$ is a lattice in W. Finally $\phi(L) = \phi(Z^n) \cap W$, $\phi(K) = \phi(Z^n) \cap V$, since

$$
\begin{array}{ccccc}
\phi(Z^n) & = & \phi(L) & + & \phi(K) \\
 & & \cap & & \cap \\
 & & W & & V
\end{array}
$$

❏

Appendix II
De Bruijn's generalized functions

In this appendix we introduce the powerful theory of generalized functions developed by de Bruijn (1973) which is the basis for several important papers on the Fourier transforms of Dirac combs (de Bruijn 1986b, 1987b). This theory is not as widely known as others, such as Schwartz distributions, although it is both powerful and elementary (it makes almost no use of measure theory). It seems appropriate to give a brief, expository introduction here, at least enough to describe the Poisson summation formula and to discuss the Wiener diagram. We will consider only functions of a single complex variable; everything we say can easily be generalized to higher dimensions. Further details as well as proofs of theorems and propositions can be found in the papers cited in this appendix.

1. Smooth functions Every theory of generalized functions is constructed with respect to a class of especially well-behaved 'ordinary' functions. In de Bruijn's theory, these functions are called *smooth*.

Definition AII.1 A complex function f, analytic in the whole complex plane, is smooth if there are positive numbers A, B, M such that

$$|f(x+iy)| \leq M \exp(-Ax^2 + By^2)$$

for all real x, y.

The set of all smooth functions is called S. The set S is a linear space: if $f, g \in S$ then $af + bg \in S$ for any scalars a and b. Indeed, it is a metric space, with inner product

$$(f, g) = \int_{-\infty}^{\infty} f(x)\overline{g(x)}dx$$

and norm

$$\|f\| = (f, f)^{1/2}.$$

Convergence of sequences and series in S is defined in the usual way. S is also closed under Fourier transformation: if $f \in S$ then $\hat{f} \in S$, where

$$\hat{f}(s) = \int_{-\infty}^{\infty} f(x) \exp(-2\pi i x s)dx \in S.$$

Thus Fourier transformation is an operator on S.

The functions in S are so well behaved, de Bruijn notes, that 'we can afford to do analysis almost in the carefree style of eighteenth century mathematics', when most questions of convergence were cheerfully ignored. Generalized functions are so devised that this carefree style can be extended to a much larger class of functions.

2. *The operators N_α.* The key element of de Bruijn's theory is a semigroup of operators on S, $\{N_\alpha\}$, parameterized by the positive reals and satisfying

$$N_\alpha N_\beta = N_{\alpha+\beta}. \tag{AII.1}$$

The N_α are integral operators:

$$(N_\alpha f)(z) = \int_{-\infty}^{\infty} K_\alpha(z, t) f(t) dt \tag{AII.2}$$

where

$$K_\alpha(z, t) = (\sinh \alpha)^{-1/2} \exp\left(\frac{-\pi}{\sinh \alpha}\left((z^2 + t^2)\cosh \alpha - 2zt\right)\right).$$

(This rather strange-looking kernel does not come out of thin air; it plays an important role in the theory of the Hermite functions and de Bruijn's theory of generalized functions can be set in that context.)

These operators commute with Fourier transformation:

Proposition AII.1 $N_\alpha \mathcal{F} = \mathcal{F} N_\alpha$.

3. *Generalized functions.* To introduce the idea of a generalized function, we first note that any $f \in S$ defines a mapping

$$F : R^+ \to S \quad \text{where} \quad F(\alpha) = N_\alpha f.$$

This mapping has the property that

$$N_\beta F(\alpha) = F(\alpha + \beta), \text{ for all } \alpha > 0, \beta > 0. \tag{AII.3}$$

Any F satisfying (AII.3) is called a *trace*. When F is defined by some $f \in S$, we call it 'the trace of f'. But (AII.3) is much more general: not every trace is the trace of a function in S.

Definition AII.2 A generalized function is a trace.

Proposition AII.2 Let F and G be generalized functions. If $F(\beta) = G(\beta)$ for any $\beta > 0$, then $F = G$.

The set of all generalized functions is called S^*. It is also a linear space. There is no inner product on S^*, but there is a kind of inner product of elements of S^* and elements of S. Let $g = N_\alpha h$ for some α (it can be proved that such an h exists; it need not be smooth but it must belong to the set S^+ defined below). Then we define

$$(F, g) = (F(\alpha), h) . \tag{AII.3}$$

The right hand side is independent of the choice of α.

Examples of generalized functions

(i) Every function in S can be embedded in S^* by the rule

$$\text{emb} f : \alpha \to N_\alpha f;$$

that this is a trace follows directly from (AII.1).Thus $\text{emb}(S) \subset S^*$. The inclusion is proper:

$$\text{emb}(S) \neq S^*.$$

(ii) The map $F(\alpha) = K_\alpha(b, t)$ is a generalized function.

Notice that when $z = b$, a real number, then the right-hand side of (AII.2) is the inner product of $K_\alpha(b, t)$ with $f \in S$ and the left-hand side becomes $(N_\alpha f)(b)$. This means that for $g = N_\alpha f$ we have

$$(K_\alpha(b, t), g) = g(b).$$

Thus $K_\alpha(b, t)$ plays the role of the Dirac delta 'at b' in this theory. We will denote it by δ_b.

Let S^+ be the set of functions such that for every $\epsilon > 0$ there is an $M > 0$ such that

$$\int_{-x}^{x} |f(t)| dt \leq M \exp(\epsilon x^2)$$

for all x; these functions are not necessarily smooth but it can be shown that $N_\alpha f(z) \in S$; for this reason, the operators N_α are called *smoothing operators*. These 'smoothed' functions can be embedded in S^* in a way completely analogous to the embedding of smooth functions defined by (AII.3), so we can speak of the embedding of S^+ in S^*. The set S^+ is important because it *contains the wave functions*. It also contains the squared integrable functions, so the Hilbert space L_2 can be embedded in S^*.

If $f \in S^+$, $g \in S$ then $(\text{emb} f, g) = (f, g)$.

Any operator on S that commutes with the N_αs can be extended to S^*. In particular, S^* is closed under Fourier transformation, and

$$\mathcal{F} F : \alpha \longrightarrow \mathcal{F} F(\alpha).$$

Thus the Fourier transform of δ_b is the map

$$\alpha \to \int_{-\infty}^{\infty} K_\alpha(t, b) \exp(-2\pi i t x) dt = \text{emb}(\exp(-2\pi i b x)).$$

The product (F, h) satisfies a 'Parseval formula':

Proposition AII.3 $(\mathcal{F} F, \mathcal{F} g) = (F, g)$ for $F \in S^*, g \in S$.

4. Dirac combs

Definition AII.3 A sequence of generalized functions $\{F_n\}$ is said to converge to a generalized function F if, for every $g \in S$, the sequence of inner products (F_n, g) converges to (F, g). A series of generalized functions $\sum F_n$ is said to converge (absolutely) to a generalized function if the sequence of partial sums converges (absolutely).

Theorem AII.1 Let $x_1, x_2, \ldots$ be a sequence of real numbers and $c_1, c_2, \ldots$ a sequence of complex numbers. If the sum $\sum c_n \exp(\epsilon x_n^2)$ converges for every $\epsilon > 0$ then $\sum c_n \delta_{x_n}$ is absolutely S^* convergent.

The conditions of the theorem are satisfied when $\{x_n\}$ is a Delone set and $c_n = 1$ for all n; hence all density 'functions' of Delone sets (see Chapter 3) are elements of S^*. In fact, the set D of absolutely convergent Dirac combs is a linear subspace of S^*.

The Fourier transform of a Dirac comb is, as we would expect, a sum of exponentials:

$$\mathcal{F}\left(\sum_{x_n \in \Lambda} \delta_{x_n}\right) = \sum_{x_n \in \Lambda} \exp(-2\pi i x_n s)$$

(where we write $\exp(u)$ for $\mathrm{emb}(\exp(u))$).

The Fourier transform of $\sum \delta_n$ is $\sum \exp(-2\pi i n x)$ It follows from Parseval's formula that

Theorem 8.6 (Poisson summation formula) $\mathcal{F}\left(\sum \delta_n\right) = \sum \delta_n$.

Using appropriately generalized forms of Theorem 3.1, we have

Corollary AII.1 The Fourier transform of $\sum_{n=-\infty}^{\infty} \delta_{nd}$ is $\sum_{n=-\infty}^{\infty} \delta_{n/d}$.

Finally, from (AII.3) we obtain

Corollary AII.2 For every $g \in S$,

$$\sum_{n=-\infty}^{\infty} g(n) = \sum_{n=-\infty}^{\infty} \hat{g}(n).$$

(This corollary is also known as the Poisson summation formula.)

5. *Concerning the Wiener diagram ...* The Wiener diagram (Chapter 3) is valid for densities of *finite* Delone sets, but not for Dirac combs. We will briefly explain why.

In order for a Wiener diagram (Chapter 3) to be valid for Dirac combs, we must first be able to define the autocorrelation functions and squared moduli of generalized functions. Convolution and multiplication are not defined for all elements of S^*, but they can be defined for certain subsets. That is, there is a set $C \subset S^*$ such that, if $F \in C$ and $G \in S^*$, the convolution product $F * G$ can be given a suitable meaning (Janssen, 1979). Moreover, there is a subset $\mathcal{M}$ on which a (rather complicated) form of multiplication can be defined. However, $\mathcal{M}$ has a simple description:

Definition AII.4 $F \in \mathcal{M}$ if F is the embedding of an analytic g satisfying, for all $\epsilon > 0$,

$$g(z) \exp(-\pi \epsilon z^2) \in S.$$

We have

Proposition AII.4 $F \in C \Longleftrightarrow \mathcal{F}F \in \mathcal{M}$

and

Proposition AII.5 $\mathcal{F}(F * G) = \mathcal{F}(F)\mathcal{F}(G)$.

Janssen showed that the Dirac delta is an element of $\mathcal{C}$, and since $\mathcal{C}$ is a linear space so are all finite sums of deltas. Thus by Proposition AII.5, these finite sums satisfy a Wiener diagram. Let $\rho_n(x)$ be such a sum and $\lim_{n\to\infty} \rho_n(x) = \rho(x)$; if $\rho \in \mathcal{C}$, then we can also show that

$$\lim_{n\to\infty} \rho_n(x) * \rho_n(x) = \rho(x) * \rho(x).$$

But it is at this point that the Wiener diagram breaks down: $\mathcal{C}$ contains only those Dirac combs whose coefficients decrease rapidly. This is certainly not the case when all the coefficients are equal to one. Thus the diagram is not even valid for $\sum \delta_n$. It is also easy to see that this generalized function, which is its own Fourier transform, is not in $\mathcal{M}$.

References

The numbers in brackets indicate the chapter(s) in which the article or book is cited.

I. Aksay, E. Baer, M. Sarikaya, and D. Tirrell (1992), Hierarchically Structured Materials, *Materials Research Society Symposium Proceedings*, Vol. 255, Pittsburgh. (See also *The New York Times*, Tuesday August 31, 1993, p. C1.) **[5]**.

J.-P. Allouche (1987), Automates finis et théorie des nombres, *Expositiones Mathematicae*, Vol. 5, 239–66. **[4]**.

J.-P. Allouche and M. Mendes France (1986), Quasicrystal Ising chain and automata theory, *Journal of Statistical Physics*, Vol. 42, Nos. 5/6, 809–21. **[4]**.

R. Ammann, B. Grünbaum, and G.C. Shephard (1992), Aperiodic tiles, *Discrete and Computational Geometry*, Vol. 8, No. 1, 1–25. **[7]**.

S. Aubry and C. Godrèche (1986), Incommensurate structure with no average lattice: an example of one-dimensional quasicrystal, *Journal de Physique, Colloque C3*, supplement No. 7, Vol 47, 187–96. **[4]**.

S. Aubry, C. Godrèche, and J.M. Luck (1988), Scaling properties of a structure between quasi-periodic and random, *Journal of Statistical Physics*, Vol. 51, 1033–75. **[4]**.

M. Baake, D. Joseph, P. Kramer, and M. Schlottmann (1990), Root lattices and quasicrystals, *Journal of Physics A: Mathematical and General*, Vol. 23, L1037–41. **[2]**.

M. Baake, P. Kramer, M. Schlottmann, and D. Zeidler (1990), Planar patterns with five fold symmetry as sections of periodic structures in 4-space, *International Journal of Modern Physics B*, 2217–68. **[2]**.

M. Baake, M. Schlottmann, and P. Jarvis (1992), Quasiperiodic tilings with tenfold symmetry and equivalence with respect to local derivability, *Journal of Physics A*, Vol. 24, 4637–54. **[7]**.

W. W. R. Ball and H. S. M. Coxeter (1974), *Mathematical Recreations and Essays*, 12th edition, University of Toronto Press, Toronto. **[4]**.

W. Barlow and H. Miers (1901), The structure of crystals, I. in *Report of the Meeting of the British Association for the Advancement of Science*, 71st meeting, London, 297–337. **[1]**.

F. P. M. Beenker (1982), Algebraic theory of non-periodic tilings of the plane by two simple building blocks: a square and a rhombus, *TH - Report 82-WSK04*, Eindhoven University of Technology. **[5, 7]**.

272

D. Berend and C. Radin (1993), Are there chaotic tilings?, *Communications in Mathematical Physics*, Vol. 152, 215–19. **[8]**.

R. Berger (1966), *Memoirs of the American Mathematical Society* No. 66, American Mathematical Society, Providence. **[5]**.

S. Berman and R. V. Moody (1994), The algebraic theory of quasicrystals with 5-fold symmetries, *Journal of Physics A: Mathematical and General*, Vol. 27, 115–29. **[6]**.

L. Bieberbach (1910), Über die Bewegungsgruppen der *n*-dimensional en Euklidischen Räume mit einem endlichen Fundamental bereich, *Göttinger Nachrichten*, 75–84. **[1]**.

L. Bieberbach (1912), Über die Bewegungsgruppen der Euklidischen Räume, *Mathematische Annallen*, Vol. 72, 400–12. **[1]**.

H. Bohr (1924), Zur theorie der fastperiodischen Funktionen I, *Acta Mathematica, Vol. 45, 29–127.* **[1]** .

H. Bohr (1925), Zur theorie der fastperiodischen Funktionen II, *Acta Mathematica*, Vol. 46, 101–214. **[1]** .

H. Bohr (1933), *Almost Periodic Functions*, Julius Springer, Berlin; Engish translation Chelsea Publishing Co., 1947. **[1]**.

J. Bohsung and H. -R. Trebin (1989), Defects in quasicrystals, in *Introduction to the Mathematics of Quasicrystals*, edited by Marko Jaríc, Academic Press, San Diego, 183–221. **[7]** .

E. Bombieri and J. Taylor (1987), Quasicrystals, tilings, and algebraic number theory: some preliminary connections, *Contemporary Mathematics*, Vol. 64, 241–64. **[3, 4]**.

N. Bourbaki (1966), *Eléments de Mathématique, Topologie Générale*, Hermann, Paris.

A. Bravais (1850), Mémoire sur les systèmes formés par des points distribués régulièrement sur un plan ou dans l'espace, *Journal de l'École Polytechnique*, Vol. XIX, 1–128, Paris. **[1]**.

A. Bravais (1851), Études Cristallographiques, *Journal de l'École Polytechnique*, Vol. XX, 101–276, Paris. **[1]**.

H. Brown, R. Bülow, J. Neubüser, H. Wondratschek, and H. Zassenhaus (1978), *Crystallographic Groups of Four Dimensional Space*, Wiley, New York. **[1]**.

N. G. de Bruijn (1973), A theory of generalized functions, with applications to Wigner distribution and Weyl correspondence, *Nieuw Archief voor Wiskunke*, Vol. 3 (XXI), 205–80. **[3, A2]**.

N. G. de Bruijn (1981a), Algebraic theory of Penrose's non-periodic tilings of the plane, *Proceedings of the Koninklijke Nederlandse Akademie van Wetenschappen Series A*, Vol. 84 (*Indagationes Mathematicae*, Vol. 43), 38–66. **[2, 4, 5, 6]**.

N. G. de Bruijn (1981b), Sequences of zeros and ones generated by special production rules, *Proceedings of the Koninklijke Nederlandse Akademie van Wetenschappen Series A*, Vol. 48, (*Indagationes Mathematicae*, Vol. 43, 27–37). **[4]**.

N. G. de Bruijn (1986a), Dualization of multigrids *Journal de Physique, Colloque C3*, supplement No. 7, Vol. 47, 85–94. **[5]**.

N. G. de Bruijn (1986b), Quasicrystals and their Fourier transform, *Proceedings of the Koninklijke Nederlandse Akademie van Wetenschappen Series A*, Vol. 89 (*Indagationes Mathematicae*, Vol. 48), 123–52. **[3, 4, 8, A2]**.

N. G. de Bruijn (1987a) A riffle-shuffle card trick and its relation to quasicrystal theory, *Nieuw Archief voor Wiskunde* Vol. 4 (5) 285–301. **[5]**.

N. G. de Bruijn (1987b), Modulated Quasicrystals, *Proceedings of the Koninklijke Nederlandse Akademie van Wetenschappen Series A*, Vol. 90 (*Indagationes Mathematicae* Vol. 49), 121–32. [3, A2].

N. G. de Bruijn (1989a), Updown generation of Beatty sequences, *Proceedings of the Koninklijke Nederlandse Akademie van Wetenschappen Series A*, Vol. 92 (*Indagationes Mathematicae* 51), 385–407. [4, 5].

N. G. de Bruijn (1989b), Remarks on Beenker patterns (preprint). [7].

N. G. de Bruijn (1990), Updown generation of Penrose tilings, *Indagationes Mathematicae, New Series*, Vol. 1 (2) 201–19. [5, 6].

N. G. de Bruijn (1992), Penrose patterns are almost entirely determined by two points, *Discrete Mathematics*, 106/107, 97–104. [6].

N. G. de Bruijn (1994), Remarks on Penrose tilings, to appear in *The Mathematics of Paul Erdös*, Springer-Verlag. [6].

N. G. de Bruijn and M. Senechal (1995), On the uniform distribution of the projection of a lattice onto a subspace, in preparation. [2] .

J. J. Burckhardt (1971), Die Briefwechsel von E. S. von Fedorov und A. Schoenflies, 1889-1908, *Archive for History of Exact Sciences*, Vol. 7, 91–141. [1].

H. Burkhill and B. C. Rennie (1983), Almost periodic generalized functions, *Mathematical Proceedings of the Cambridge Philosophical Society*, Vol. 94, 149–66. [3] .

S. E. Burkov (1988), Absence of weak local rules for the planar quasicrystalline tiling with 8-fold symmetry, *Communications in Mathematical Physics* Vol. 119, 667–75. [7].

S. E. Burkov (1992), Modeling decagonal quasicrystals: random assembly of interpenetrating decagonal clusters, *Journal de Physique I France*, Vol. 2, 695–706. [6].

J. W. S. Cassells (1965), *Diophantine Approximation*, Cambridge Tracts in Mathematics and Mathematical Physics No. 45, Cambridge University Press. [4, A1].

D. C. Champeney (1987), *A Handbook of Fourier Transforms*, Cambridge University Press. [3].

K. Chandrasekharan (1989), *Classical Fourier Transforms*, Springer-Verlag, New York Berlin Heidelberg. [3].

J. H. Conway and N. J. A. Sloane (1988), *Sphere Packings, Lattices,and Groups*, Springer Verlag, New York Berlin Heidelberg. [2].

A. Córdoba (1988), La formule sommatoire de Poisson, *Comptes Rendus de L'Academie des Sciences de Paris*, Vol. 306, Sèrie I, 373–76. [3].

J. M. Cowley (1981), *Diffraction Physics*, North-Holland, Amsterdam, 2nd edition (paperback 1986). [3].

H. S. M. Coxeter (1969), *Introduction to Geometry*, New York, Wiley and Sons. [4, 5] .

H. S. M. Coxeter (1973) *Regular Polytopes*, Dover, New York, 1973. [2].

H. S. M. Coxeter (1993), Cyclotomic integers, nondiscrete tessellations, and quasicrystals, *Indagationes Mathematicae New Series*, Vol. 4, No. 1, 27–8. [6].

L. Danzer (1989), Three dimensional analogues of the planar Penrose tilings and quasicrystals, *Discrete Mathematics*, Vol. 76, 1–7. [7].

L. Danzer (1991), Quasiperiodicity: local and global aspects, preprint, University of Bielefeld, ZiF-Nr. 6. [1].

L. Danzer (1993b), A single prototile, which tiles space, but neither periodically nor quasiperiodically, preprint. [7].

L. Danzer, Z. Papadopolos, and A. Talis (1993), Full equivalence between Socolar's tilings and the (A,B,C,D)-tilings leading to a rather natural decoration, *International Journal of Modern Physics B*, Vol. 107, Nos. 6 and 7, 1379–86. **[7]**.

M. Dekking, M. Mendes France, and A. van der Poorten [1982a], Folds!, *The Mathematical Intelligencer*, Vol. 4, No. 3, 130–8. **[4]**.

M. Dekking, M. Mendes France, and A. van der Poorten [1982b], Folds! II: symmetry disturbed, *The Mathematical Intelligencer*, Vol. 4, No. 4, 173–95. **[4]**.

B. N. Delone, A. D. Aleksandrov, and N. N. Padurov (1934), *Foundations of the Structural Analysis of Crystals*, Leningrad-Moscow, O.N.T.I. (in Russian). **[1]**.

B. N. Delone (Delaunay) (1929), Sur la partition régulière de l'espace à 4 dimensions, *Izvestia Akademii Nauk S.S.S.R., Otdelenie Fiziko–Matematicheskikh Nauk* 79–110, 145–64. **[2]**.

B. N. Delone, N. P. Dolbilin, M. I. Shtogrin, and R. V. Galiulin (1976) A local criterion for regularity of a system of points, *Reports of the Acadamy of Sciences of the U.S.S.R. (in Russian), Vol. 227. (English translation: Soviet Math. Dokl.*, 17 (1976), No. 2, 319–22.) **[1]**.

R. Descartes (1644), *Traité sur la lumière* **[2]**.

M. W. C. Dharma-wardana, A. H. MacDonald, D. J. Lockwood, J.-M. Baribeau, and D. C. Houghton, Raman scattering in Fibonacci superlattices, *Physical Review Letters*, Vol. 58, No. 17, 1761–4. **[4]**.

V. P. Dimitriev, Yu. M. Gufan, S. B. Rochal, and P. Tolédano (1990), Theory of the formation of quasicrystals, *Journal de Physique France*, Vol. 51, 2399–405. **[7]**.

D. DiVincenzo and P.J. Steinhardt (1991), *Quasicrystals: the state of the art*, World Scientific Publishers, Singapore. **[preface]**.

N. Dolbilin (1993), On the periodicity of tilings (preprint), **[7]**.

M. Duneau and C. Oguey (1990), Displacive transformations and quasicrystalline symmetries, *Journal de Physique France*, Vol. 51, 5–19 . **[8]**.

P. Duval (1964), *Homographies, Quaternions, and Rotations*, Oxford Mathematical Monographs, Clarendon Press **[2]**.

S. Dworkin (1993), Spectral theory and x-ray diffraction, *Journal of Mathematical Physics*, Vol. 34, No. 7, 2965–67. **[3]**.

S. Dworkin and J. I. Shieh (1993), Deceptions in quasicrystal growth, to appear in *Communications in Mathematical Physics.* **[7]**.

S. Y. Edgerton, Jr (1984), Galileo, Florentine 'disegno', and the 'strangespottednesse' of the moon, *Art Journal*, Vol. 44, 225–32. **[1]**.

S. J. L. van Eijndhoven (1987), Functional analytic characterizations of the Gelfand–Shilov Spaces S_α^β, *Indagationes Mathematicae A*, Vol. 90, No. 2, 133–44. **[3, 8]**.

V. Elser (1986), The diffraction pattern of projected structures, *Acta Crytallographica A*, Vol. 42, 36–43. **[4]**.

V. Elser and N. J. A. Sloane (1987), A highly symmetric four-dimensional quasicrystal, *Journal of Physics A: Mathematical and General*, Vol. 20, 6161–7. **[2]**.

P. Engel (1986), *Geometric Crystallography*, Reidel, Dordrecht. **[1]**.

A. C. D. van Enter and J. Miekisz (1992), How should one define a (weak) crystal?, Journal of Statistical Physics, Vol. 66, 1147–53. **[4]**.

E. S. Fedorov (1885), Nachala Ucheniya o Figurakh, *Notices of the Imperial St*

Petersburg Mineralogical Society, 2nd series, Vol. 21, 1–279. **[1]**.

E. S. Fedorov (1891), Symmetry of regular systems of figures (in Russian), *Notices of the Imperial St Petersburg Mineralogical Society*, Vol. 28, 1–146. English translation by David and Katharine Harker, ACA Monograph No. 7. **[1]**.

E. S. Fedorov (1920), *Das Kristallreich: Tabellen zur Kristallochemischen Analyse*, Academy of Sciences, St Petersburg **[1]**.

A. Fontaine (1991), An infinite number of plane figures with Heesch number two, *Journal of Combinatorial Theory, Series A*, Vol. 57, 151–6. **[5]**.

G. Frobenius (1911), Über die unzerlegbaren diskreten Bewegungsgruppen, *Sitzungsberichte der Koninglichen Preussischen Akademie der Wissenschaften, Berlin*, 654–66. **[1]**.

F. Gähler (1994), Binary tiling quasicrystals and matching rules, preprint **[7]**.

F. Gähler and J. Rhyner (1986), Equivalence of the generalized grid and projection methods for the construction of quasiperiodic tilings, *Journal of Physics A: Mathematical and General*, Vol. 19, 267–77. **[5]**.

I. M. Gelfand and G. E. Shilov (1968), *Generalized Functions*, Vol. 2, Academic Press, New York. **[8]**.

O. Glaser and D. Wrinch (1953), Diffraction patterns in nineteenth-century astronomy and twentieth-century x-ray crystallography, in *Science, Medicine and History*, Oxford University Press. **[3]**.

L. Glass, A. Goldberger, and J. Belair (1986), Dynamics of pure parasystole, *American Journal of Physiology*, Vol. 251, H841–7. **[4]**.

C. Godrèche (1989), The sphinx: a limit-periodic tiling of the plane. *Journal of Physics A: Mathematical and General*, Vol. 22, 1163–6. **[8]**.

C. Godrèche (1990), Types of order and diffraction spectra for tilings of the line, in *Number Theory and Physics*, Proceedings of the Winter School, Les Houches, France, 7–16 March, 1989, edited by J.-M. Luck, P. Moussa, and M. Waldschmidt, Springer-Verlag, Berlin-Heidelberg. **[4]**.

C. Godrèche (1991), Non Pisot tilings and singular scattering, *Phase Transitions*, Vol. 32, 45–79. **[8]**.

C. Godrèche and F. Lançon (1992), A simple example of a non-Pisot tiling with five-fold symmetry, *Journal de Physique*, Vol. 2, 207–20. **[6, 8]**.

C. Godrèche and J.-M. Luck (1989), Quasiperiodicity and randomness in tilings of the plane, *Journal of Statistical Physics*, Vol. 55, No. 1, 1–28. **[6,8]**.

C. Godrèche and J.-M. Luck (1992), Indexing the diffraction spectrum of a non-Pisot self-similar structure, *Physical Review B*, Vol. 45, No. 1, 176–85. **[4]**.

C. Godrèche, J.-M. Luck, A. Janner and T. Janssen (1993), Fractal atomic surfaces of self-similar quasiperiodic tilings of the plane, *Journal de Physique I France*, Vol. 3, 1921–39. **[4, 5, 8]**.

C. Godrèche, J.-M. Luck, and F. Vallet (1987), Quasiperiodicity and types of order: a study in one dimension, *Journal of Physics A*, Vol. 20, 4483–99. **[4]**.

S. Golomb (1964), Replicating figures in the plane, *Mathematical Gazette*, Vol. 48, 403–12. **[5]**

B. Grünbaum and G. C. Shephard (1980), Tiling with congruent tiles, *Bulletin of the American Mathematical Society*, N.S., Vol. 3, 951–73. **[1]**.

B. Grünbaum and G. C. Shephard (1977), The eighty-one types of isohedral tilings in the plane, *Mathematical Proceedings of the Cambridge Philosophical Society*, Vol. 82, 177–96. **[7]**.

B. Grünbaum and G. C. Shephard (1987), *Tilings and Patterns*, W.H. Freeman, New York. **[preface, 1, 5]**.

H. Hadwiger (1940), Über ausgezeichnete Vektorsterne und regulöre Polytope,

Commentarii Mathematici Helvetici, Vol. 13, 90–107. **[2]**.

G. Harburn, C. A. Taylor, and T. R. Welberry (1975), *Atlas of Optical Transforms*, London, G. Bell & Sons, Ltd. **[3, 8]**.

G. H. Hardy and E. M. Wright (1962), *An Introduction to the Theory of Numbers*, 4th edition, Oxford University Press.**[A2]** .

G. H. Hardy and M. Riesz (1915), *The General Theory of Dirichlet's Series*, Cambridge Tracts in Mathematics and Mathematical Physics Number 18, Cambridge University Press. Reprinted 1952. **[3]** .

R. J. Haüy (1822), *Traité de Cristallographie*, 3 volumes, Paris. **[1]**.

H. Heesch (1935), Aufbau der Ebene aus kongruenten Bereichen, *Nachr. Ges. Wiss. Göttingen, New Ser.*, Vol. 1, 115–7. **[1]**.

C. Hermann (1949), Kristallographie in Räumen beliebige Dimensionszahl. I. Die Symmetrieoperationen., *Acta Crystallographica*, Vol. 2, 139–44. **[2]**.

D. Hilbert (1900), Mathematische Probleme, *Göttinger Nachrichten*, 253–97. **[1]**.

H. Hiller(1985), The crystallographic restriction in higher dimensions, *Acta Crystallographica*, 541–4. **[2]**.

E. Hlawka (1979), *The Theory of Uniform Distribution*, Bibliographisches Institut, Mannheim; A B Academic Publ., Berkhamsted (1984) **[A1]** .

A. Hof (1992), *Quasicrystals, Aperiodicity, and Lattice Systems*, Thesis, Groningen, The Netherlands. **[5]** .

A. Hof (1993), On diffraction by aperiodic structures, submitted to *Communications in Mathematical Physics*. **[3, 4]**.

R. Hooke (1665), *Micrographia*, Jo. Martin and Js. Allestry, London. **[1]**.

C. Huyghens (1690), *Traité de la Lumière*, Pierre van der Aa, Leiden. **[1]**.

K. Ingersent (1991), Matching rules for quasicrystalline tilings, in *Quasicrystals: the state of the art*, edited by D. Di Vincenzo and P.J. Steinhardt, World Scientific Publishers, Singapore. **[5, 7]**.

A. Janner (1991), Which symmetry will an ideal quasicrystal admit?, *Acta Crystallographica* A47, 577–90. **[2]** .

A. Janner and T. Janssen (1979), Superspace Groups, *Physica 99A*, 47–76. **[1]**.

C. Janot (1993), *Quasicrystals: a primer*, Oxford, Oxford University Press. **[preface]**.

A. J. E. M. Janssen (1979), Convolution theory in a space of generalized functions, *Proceedings of the Koninklijke Nederlandse Akademievan Wetenschappen, Series A*, Vol. 82, No. 3, 283–305. **[A2]** .

C. Jordan (1868), Mémoire sur les groupes de mouvements, *Annali de matematica pura ed applicata*, Ser. II, Vol. II, No. 3, 167–215, 322–45. **[1]**.

P. A. Kalugin, A. Yu. Kitaev, and L. S. Levitov (1986), 6-dimensional properties of $Al_{0.86}Mn_{0.14}$ alloy, *Journal de Physique Lettres*, Vol. 46, 601–7. **[2]**.

A. Katz (1988), Theory of matching rules for the 3-dimensional Penrose tiling, *Communications in Mathematical Physics*, Vol. 118, 263–88. **[7]**.

A. Katz (1989), Some local properties of the 3-dimensional Penrose tilings, in *Introduction to the Mathematics of Quasicrystals*, edited by Marko Jarić, Academic Press, San Diego, CA, 147–82. .

A. Katz and M. Duneau (1986), Quasiperiodic patterns and icosahedral symmetry, *Journal de Physique*, Vol. 47, 181–96. **[2, 4]**.

J. Kepler (1611), *Strena seu de nive sexangula*, Francofurti ad Moenum. **[1]**.

J. Kepler (1619), *Harmonices Mundi*, Book II. (See also the English translation of the relevant parts by J. V. Field, Kepler's star polyhedra, in Vistas in Astronomy, Vol. 23, 1979, 109–41.) **[1, 6]**.

H. Kesten (1966), On a conjecture of Erdös and Szüsz related to uniform distribution mod 1, *Acta Arithmetica*, Vol. 12, 193–212. **[7]**.

M. Klèman and A. Pavlovitch (1987), Generalized 2D Penrose tilings: structural properties, *Journal of Physics A: Mathematical and General*, Vol. 20, 687–702. **[7]** .

M. Klèman (1992), Dislocations and disvections in aperiodic crystals, Journal de Physique I France, Vol. 2 69–87. **[7]**.

R. Klitzing, M. Schlottmann, and M. Baake, Perfect matching rules for undecorated triangular tilings with 10-, 12-, and 8-fold symmetry, International Journal of Modern Physics, Vol. 7, Nos. 6 and 7, 1453–73. **[7]**.

V. E. Korepin, F. Gähler, and J. Rhyner (1988), Quasi-periodic tilings: a generalized grid-projection method, *Acta Crystallographica A*, Vol. 44, 667–72. **[5]**.

P. Kramer and R. W. Haase (1989), Group theory of icosahedral crystals, in *Introduction to the Mathematics of Quasicrystals*, edited by Marko Jarić, Academic Press, San Diego, pp. 81–146. **[2]** .

P. Kramer and R. Neri (1984), On periodic and non-periodic space fillings of E^m obtained by projection, *Acta Crytallographica A*, Vol. 40, 580–7. **[2, 7]**.

T. Kuhn (1970), *The Structure of Scientific Revolutions*, University of Chicago Press. **[1]**.

M. La Brecque (1987/88), Opening the door to forbidden symmetries, *Mosaic*, Vol. 18, No. 4, Winter, 2–23. **[1]**.

F. Lançon and L. Billard (1988), Two-dimensional system with a quasi-crystalline ground state, *Journal de Physique France*, Vol. 49, 249–56. **[6, 8]**.

F. Lançon and L. Billard (1993), Binary tilings and quasi-quasicrystalline tilings, *Phase Transitions*, Vol. 44, 37–46. **[4, 7]**.

E. Lerner (1988), How to grow a quasicrystal, *IBM Research Magazine*, winter 1988, 8–11. **[7]** .

T. Q. T. Le (1993), Local rules for pentagonal quasicrystals, to appear in *Discrete and Computational Geometry*. **[6, 7]**.

T. Q. T. Le and S. Piunikhin (1993), Local rules for multidimensional quasicrystals, to appear in *Differential Geometry and its Applications*. **[7]** .

T. Q. T. Le, S. Piunikhin, and V. Sadov (1992a), Local rules for quasiperiodic tilings of quadratic 2-planes in R^4, *Communications in Mathematical Physics*, Vol. 150, 23–44. **[7]**.

T. Q. T. Le, S. Piunikhin, and V. Sadov (1992b), Geometry of quasicrystals, *Russian Mathematical Surveys* (in Russian), Vol. 6, 41–102. **[7]**.

D. Levine and P. J. Steinhardt (1986), Quasicrystals I. Definition and structure, *Physical Review B*, Vol. 34, no. 2, 596–615. **[2, 7]**.

M. J. Lighthill (1958), *An Introduction to Fourier Analysis and Generalised Functions*, Cambridge University Press. **[3]**.

C. Linneaus (1768), *Systema Naturae*, Vol. 3, Laurenti Salvii, Stockholm. **[1]**.

S. Lipson and H. Lipson (1981), *Optical Physics*, Cambridge University Press, 2nd edition. **[3]**.

S. Y. Litvin, A. B. Romberger, and D. B. Litvin (1988), Generation and experimental measurement of a one-dimensional quasicrystal diffraction pattern, *Am. J. Phys.*, Vol. 56, No. 1, 72–5. **[4]** .

J.-M. Luck, C. Godrèche, A. Janner and T. Janssen (1993), The nature of the atomic surfaces of quasiperiodic self-similar structures, *Journal of Physics A*, Vol. 26, 1951–99. **[4]**.

R. Lück (1993), Basic ideas of Ammann bar grids, *International Journal of Modern Physics*, Vol. 7, Nos. 6 and 7, 1437–53. **[7]**.

F. Lunnon and P. Pleasants (1987a), Quasicrystallographic tilings, *Journal de mathématiques pures et appliqués*, Vol. 66, 217–63. **[4, 5]**.

F. Lunnon and P. Pleasants (1992), Characterization of two-distance sequences, *Journal of the Australian Mathematical Society*, Series A, Vol. 53, 198–218. **[4]**.

A. Mackay (1982), Crystallography and the Penrose pattern, *Physica A*, Vol. 114, 609–13. **[8]** .

K. Mahler (1926), The spectrum of an array, *Journal of Mathematics and Physics*, Vol. VI, 158–63. **[4]**.

M. Mendes France (1984), Folding paper and thermodynamics, *Physics Reports*, Vol. 103, Nos. 1–4, 161–72. **[4]**.

M. Mendes France (1988), Some applications of the theory of automata, in *Proceedings of Prospects of the Mathematical Sciences*, World Science Publishers, 127–40. **4]**.

M. Mendes France and A. J. van der Poorten (1981), Arithmetic and Analytic Properties of Paper Folding Sequences, *Bulletin of the Australian Mathematical Society*, Vol. 24, 123–31. **[4]**.

M. Mendes France and J. O. Shallit (1989), Wire Bending, *Journal of Combinatorial Theory, Series A*, Vol. 50, No. 1, 1–23. **[4]**.

D. Mermin (1991), (Quasi)crystallography is better in Fourier space, in *Quasicrystals: the state of the art*, edited by D. DiVincenzo and P. J. Steinhardt, World Scientific Publishers, Singapore, 133–83. **[1, 2]**.

D. Mermin (1992), The space groups of icosahedral quasicrystals and cubic, orthorhombic, monoclinic, and triclinic crystals, *Reviews of Modern Physics* No. 396, Vol. 64, No. 1, 3–49. **[1]**.

H. Metzger (1922), *La Genese de la Science des Cristaux*, News Funch, Paris. (Also published (1969) by Librairie Scientifique et Technique, Albert Blanchard, Paris.) **[1]**.

Y. Meyer (1972), *Algebraic Numbers and Harmonic Analysis*, North-Holland, Amsterdam. **[2]** .

L. Michel and J. Mozryzmas (1989), Les concepts fondamentaux de la cristallographie, *Comptes Rendus de L'Academie des Sciences de Paris*, Vol. 308, Sèrie II, 151–8. **[1]**.

H. Minkowski (1907), *Diophantische Approximationen*. **[1]**.

R. V. Moody and J. Patera (1992), Voronoï and Delaunay cells of root lattices: classification of their faces and facets by Coxeter–Dynkin diagrams, *Journal of Physics A: Mathematical and General*, Vol. 25, 5089–134. **[2]**.

R. V. Moody and J. Patera (1993), Quasicrystals and Icosians, *Journal of Physics A: Mathematical and General*, Vol. 26, 2829–53. **[2]**.

H.-U. Nissen [1990], A two-dimensional quasiperiodic dodecagonal tiling by two pentagons, in *Quasicrystals, Networks, and Molecules of Five-fold Symmetry*, I. Hargittai, ed., VCH, Weinheim,181–99. **[7]**.

C. Oguey, M. Duneau, and A. Katz (1988), A geometrical approach to quasiperiodic tilings, *Communications in Mathematical Physics*, Vol. 118, 99–118. **[5, 7]**.

A. Okabe, B. Boots, and K. Sugihara (1992), *Spatial Tessellations: concepts and applications of Voronoï diagrams*, Chichester, John Wiley and Sons, Ltd. **[2]**.

A. L. Patterson (1962), Experiences in crystallography–1924 to date, in *Fifty Years of X-Ray Diffraction*, P.P. Ewald, editor, Oosthoek, Utrecht. **[3]**.

R. Penrose (1974), Pentaplexity, *Bulletin of the Institute for Mathematics and Applications*, Vol. 10, 266–271. **[6]**.

R. Penrose (1989), Tilings and quasicrystals: a non-local growth problem, in *Introduction to the Mathematics of Quasicrystals*, edited by Marko Jarić, Academic Press, San Diego, CA, 53–80. **[7]**.

C. Pisot (1946), Répartition (mod 1) des puissances successives des nombres réels, *Commentarii Mathematici Helvetici*, Vol. 19, 153–60. **[4]**.

H. Porta and K. B. Stolarsky (1990), Half-silvered mirrors and Wythoff's game, *Canadian Mathematical Bulletin*, Vol. 33, No. 1, 119–25. **[4]**.

R. Porter (1988), The Applications of the Properties of Fourier Transforms to Quasicrystals, MSc. Thesis, Rutgers University. **[4]**.

M. Queffélec (1987), *Substitution Dynamical Systems–Spectral Analysis*, Lecture Notes in Mathematics, No. 1294, Springer-Verlag. **[4]**.

C. Radin (1991), Global order from local sources, *Bulletin of the American Mathematical Society*, New Series, Vol. 25, No. 2, 335–364. **[8]**.

C. Radin (1992), Z^n vs. Z actions for systems of finite type, *Contemporary Mathematics*, Vol. 135, 339–42. **[5]** .

C. Radin and M. Wolff (1992), Space tilings and local isomorphism, *Geometria Dedicata*, Vol. 42, 355–60. **[5, 7]**.

C. Radin (1993a), Symmetry of tilings of the plane, *Bull. Amer. Math. Soc*, Vol. 29, 213–7. **[7]**.

C. Radin (1993b), Aperiodic tilings in higher dimensions, to appear in *Proceedings of the American Mathematical Society*. **[7]**.

C. Radin (1994), Pinwheel tilings of the plane, to appear in *Annals of Mathematics*. **[7]**.

K. Reinhardt (1928), Zur Zerlegung der euklidischen Räume in kongruente Polytope, *Sitzungsberichte der Preuss. Akademie der Wissenschaften Berlin*, 150–5. **[1]**.

B. C. Rennie (1982), On generalized functions, *Journal of Applied Probability*, Vol. 19, 139–56. **[3]**.

J. Richter-Gebert (1993), Line arrangements and zonotopal tilings: a little printer exercise, *Hyperspace*, Vol. 2, no. 3, Japan Institute of Hyperspace Science, 8–17. **[8]** .

A. Robinson (1993), The dynamical theory of tilings and quasicrystallography, preprint. **[5, 6]**.

D. Ruelle (1970), Integral representation of states on a C^* algebra, *Journal of Functional Analysis*, Vol. 6, No. 1, 116–51. **[3]** .

D. Ruelle (1982), Do turbulent crystals exist?, *Physica*, Vol. 113A 619–23. **[8]**.

O. P. Scherbak (1988), Wavefronts and reflection groups, *Russian Mathematical Surveys*, Vol. 43, No. 3, 149–94. **[2]** .

D. Shechtman, I. Blech, D. Gratias, and J.W. Cahn, (1984), Metallic phase with long-range orientational order and no translational symmetry, *Physical Review Letters*, Vol. 53, 1951–3. **[1]**.

M. Schlottmann (1993a), *Geometrische Eigenschaftern quasiperiodischer struckturen*, thesis, University of Tübingen, Germany.

M. Schlottmann (1993b), Periodic and quasiperiodic Laguerre tilings, International Journal of Modern Physics, Vol. 7, Nos. 6 and 7, 1351–63. **[7]**.

P. Schmitt (1988), An aperiodic prototile in space, informal note. **[7]**.

A. Schoenflies, *Kristallsystem und Kristallstruktur*, Leipzig, Teubner, 1891; Springer Verlag, New York Heidelberg Berlin, 1984. **[1]**.

E. Scholtz (1989), *Symmetrie–Gruppe–Dualität*, Birkhäuser, Basel. **[1]**.

L. Schwartz (1966), *Theorie des Distributions*, Hermann, Paris. **[3]**.

R. L. E. Schwarzenberger (1980), *N-dimensionalCrystallography*, Pitman, London.

[2].
M. Senechal (1981), Which tetrahedra fill space?, *Mathematics Magazine*, Vol. 54, 227–43. **[1]**.

M. Senechal (1985), Introduction to mathematical crystallography, in *I.H.E.S. Workshop on Mathematical Crystallography*, IHES preprint P/85/47. **[preface, 1]**.

M. Senechal and R. V. Galiulin (1984), An introduction to the theory of figures: the geometry of E. S. Fedorov, *Structural Topology*, Vol. 10, 5–22. **[1]**.

M. Senechal (1986), Geometry and crystal symmetry, in *Symmetry: unifying human understanding*, edited by Istvan Hargittai, New York, Pergamon Press, 565–78. **[7]** .

M. Senechal (1990), Brief history of geometrical crystallography, in *Historical Atlas of Crystallography*, J. Lima-de-Faria, ed., published for the International Union of Crystallography by Kluwer, Dordrecht, 43–59. **[1]**.

M. Senechal (1991a), Generalizing Crystallography: puzzles and problems in dimension one, in *Quasicrystals, Networks, and Molecules of Five-fold Symmetry*, edited by Istvan Hargittai, New York, VCH Publishers, 19–33. **[4]**.

M. Senechal (1991b), *Crystalline Symmetries*, Adam Hilger, Bristol. **[2]**.

M. Senechal (1991c), Finding the finite subgroups of O(3), *American Mathematical Monthly*, Vol. 97, No. 4, 329–35. **[2]**.

M. Senechal (1992), Introduction to Lattice Geometry, in *From Number Theory to Physics*, P. Moussa and M. Waldschmidt, eds., Springer Verlag. **[2]**.

M. Senechal (1994), Calendars, pacemakers, and quasicrystals, in *In Eves' Circles*, MAA Notes No. 34, Mathematical Association of America, Washington, 117–126. **[4]**.

C. Series (1985), The geometry of Markoff numbers, *Mathematical Intelligencer*, Vol. 7, No. 3, 20–29. **[4]**.

M. I. Shtogrin, (1973), Regular Dirichlet–Voronoï Partitions, *Proceedings of the Steklov Institute of Mathematics* No. 123, Moscow. **[2]**.

J. E. S. Socolar (1989), Simple octagonal and dodecagonal quasicrystals, *Physical Review B*, Vol 39, No. 15, 10519–51. **[7]**.

J. E. S. Socolar (1990a), Weak matching rules for quasicrystals, *Communications in Mathematical Physics*, Vol. 129, 599–619. **[7]**.

J. E. S. Socolar (1990b), The alternation condition and 2D quasicrystals, in *Quasicrystals*, edited by T. Fujuwara and T. Ogawa, Springer Series in Solid-State Sciences, Vol. 93, Springer-Verlag, Berlin Heidelberg, 101–11. **[6]**.

J. E. S. Socolar (1991), Growth rules for quasicrystals, in *Quasicrystals: the state of the art*, edited by D. DiVincenzo and P. J. Steinhardt, World Scientific Publishers, Singapore, 213–38. **[7]**.

J. E. S. Socolar and P. Steinhardt (1986), Quasicrystals. II. Unit-cell configurations, *Physical Review B*, Vol. 34, 617–47. **[7]**.

B. Solomyak (1993), Dynamics of self-similar tilings, preprint. **[3]**.

P. Sonneborn, Inflation and Deflation Properties of Defect Tilings (in E^n), thesis, University of Dortmund (in preparation) **[7]**.

N. Steno (1667), *De Solido intra Solidum Naturaliter Contento*, Star, Florence. (English translation: The Prodromus of Nicolaus Steno's Disseration Concerning a Solid Body Enclosed by the Process of Nature within a Solid, J. Winter, New York, (1916).) **[1]**.

P. Stephens (1989), The icosahedral glass model, in *Extended Icosahedral*

Structures, edited by Marko Jaríc and D. Gratias, Academic Press, San Diego, CA, 37–104. **[7]** .

W. Steurer and K. H. Kuo (1990), Five-dimensional structure analysis of decagonal $Al_{65}Cu_{20}Co_{15}$, *Acta Crystallographica B*, Vol. 46, 703–12. **[preface, 2]**.

K. Strandburg (1989), Random-tiling quasicrystals, *Physical Review B*, Vol. 40, No. 9, 6071–83. **[6]** .

D. Struik (1925), Het probleem 'De Impletione Loci', *Nieuw Archief voor Wiskunde*, 121–34. **[1]**.

W. Thurston (1989), Groups, tilings, and finite state automata, Summer 1989 AMS Colloquium Lectures. Available in preprint form from The Geometry Center, 1200 Washington Avenue South, Minneapolis MN 55415; the first lecture was published as Conway's tiling groups, *The American Mathematical Monthly*, Vol. 97, No. 8, October 1990, 757–73. **[5]**.

G. Voronoï (1908), Recherches sur les paralléloèdresprimitifs, *Journal für die reine und angewandte Mathematik*, Vol. 134, 198–287. **[1]**.

G. Voronoï (1909) Part II of Voronoï (1908), *Journal für die reine und angewandte Mathematik*, Vol. 136, 67–181. **[1]**.

S. Wagon (1990), *Mathematica in Action*, W. H. Freeman, NY. **[8]**.

W. Waterhouse (1972), The discovery of the regular solids, *Archive for History of Exact Sciences*, Vol. 9, 212–21. **[1]**.

H. Weyl (1916), Über die Gleichverteilung van Zahlen mod. Eins, *Mathematicsche Annalen, Vol. 77, 313–52; reprinted in HermannWeyl, Gesammelte Abhandlungen*, Vol. I, edited by K. Chandrasekharan, Springer-Verlag, Berlin Heidelberg New York, 1968, 563–99.**[2, A1]** .

M. Widom, K. Strandburg, and R. Swendson (1987), Quasicrystal equilibrium state, *Physical Review Letters*, Vol. 58, No. 7, 706–9. **[6]**.

N. Wiener (1930), Generalized harmonic analysis, *Acta Mathematica*, Vol. 55, 117–258 (reprinted by MIT Press, 1964). **[3]**.

P. M. de Wolff, and W. van Aalst (1972), The four-dimensional space group of $\gamma - Na_2Co_3$, *Acta Crystallographica A*, Vol. 28 111– **[preface, 1]**.

Subject Index